Standards-Based Math
5-6

Written by
Alaska Hults

Editor: Collene Dobelmann
Illustrator: Corbin Hillam
Designer/Production: Moonhee Pak/Rosa Gandara
Cover Designer: Barbara Peterson
Art Director: Tom Cochrane
Project Director: Carolea Williams

Table of Contents

Introduction

Each book in the *Power Practice*™ series contains dozens of ready-to-use activity pages to provide students with skill practice. The fun activities can be used to supplement and enhance what you are already teaching in your classroom. Give an activity page to students as independent class work, or send the pages home as homework to reinforce skills taught in class. An answer key is included at the end of each book to provide verification of student responses.

Standards-Based Math 5–6 progresses from basic skills and concepts to the more complex within each section. The structure of the book enhances student learning and enables them to meet the next challenge with confidence. Students will receive reinforcement in skills from the following math strands:

- Number and Operations
- Algebra
- Geometry
- Measurement
- Data Analysis and Probability

Use these ready-to-go activities to "recharge" skill review and give students the power to succeed!

Number Value

NUMBER AND OPERATIONS

Millions	Hundred Thousands	Ten Thousands	Thousands	Hundreds	Tens	Ones	Tenths	Hundredths	Thousandths
9	3	2	6	4	0	8	1	5	7

Write the number in words.

1 462

_____

2 402

_____

3 460.3

_____

4 82,974

_____

5 50,004.08

_____

6 4,768,312

_____

Write the value of the underlined digit.

7 45,<u>7</u>65

_____

8 <u>3</u>79,250.008

_____

9 100,923.46<u>5</u>

_____

10 46,0<u>3</u>8.9

_____

Order the numbers from greatest to least.

11 546; 564; 465; 645

_____

12 5,043; 3,087; 4,503; 3,708

_____

13 43.04; 44.03; 34.04; 40.34

_____

14 234.89; 254.45; 234.98; 345.54

_____

Standards–Based Math • 5–6 © 2004 Creative Teaching Press

Number Notations

NUMBER AND OPERATIONS

Numbers can be expressed in different forms or notations.
standard =   3,254
expanded notation =   3,000 + 200 + 50 + 4
scientific notation = $(3 \times 10^3) + (2 \times 10^2) + (5 \times 10) + 4$

Write each number in expanded and scientific notation.

1 365 _____

2 9,787 _____

3 3,700 _____

4 42,899 _____

5 5,050,555 _____

Find each error. Rewrite the expanded or scientific notation correctly.

6 291 = 200 + 9 + 1 _____

7 5,798 = $(5 \times 10^4) + (7 \times 10^3) + (9 \times 10^2) + (8 \times 10^0)$_____

8 7,080 = $(7 \times 10^4) + (8 \times 10^2) + 8$ _____

9 43,256 = 40,00 + 3,000 + 200 + 50 + 6 _____

10 33,024 = 30,000 + 3,000 + 200 + 40 _____

Inverse Operations

NUMBER AND OPERATIONS

Inverse operations are useful for checking your accuracy.

67	121	121	45	27
+ 54	- 67	- 54	- 27	+ 18
121	54	67	18	45

Add or subtract. Then use the inverse operation to check your work.

1   245
     + 563

2   3,692
     + 897

3   785
     − 598

4   98,534
     + 43,877

5   23,322
     + 65,533

6   3,744
     − 1,976

7   59,043
     + 23,967

8   54,255
     − 32,738

9   567,411
     − 269,577

10   24,803
     + 86,129

11   46,003
     − 10,895

12   12,879
     + 98,324

Standards–Based Math • 5–6 © 2004 Creative Teaching Press

Riddle

NUMBER AND OPERATIONS

Solve.

1
```
   3,222
   6,060
+  7,679
```

2
```
   3,692
     491
+    897
```

3
```
   8,291
−  8,138
```

4
```
  41,258
  49,108
+ 45,867
```

5
```
  29,382
   7,239
+ 48,470
```

6
```
  34,754
− 17,796
```

7
```
  51,238
  71,995
+ 23,967
```

8
```
  54,983
− 32,794
```

9
```
  587,056
− 299,439
```

10
```
  44,853
  56,599
+ 28,426
```

11
```
  50,084
− 29,695
```

12
```
  72,279
  53,062
+ 38,354
```

Match the answers to letters from the code. Decode the message.

153 =	E	5,080 =	H	16,958 =	T
16,961 =	T	20,389 =	A	22,189 =	E
85,091 =	E	129,878 =	G	147,200 =	T
163,695 =	S	287,617 =	R	136,233 =	L

What always ends everything?

___ ___ ___   ___ ___ ___ ___ ___ ___   ___!
 1   2   3   4   5   6   7   8   9   10

Comparing Integers

NUMBER AND OPERATIONS

> **Positive integers** represent an increase of value. Integers are used most often to note a change in measurement such as an increase in weight, volume, or money.
> **Negative integers** represent a decrease of value. A loss of weight, volume, or money might be noted with a negative integer.

Circle the greatest value. Underline the least value.

1) 7, 3, ⁻8, ⁻9, 2

2) ⁻5, 6, 12, ⁻12, 4, 0

3) ⁻4, ⁻8, ⁻5, ⁻6, 0, ⁻3

4) ⁻2, ⁻9, ⁻5, ⁻3, ⁻13

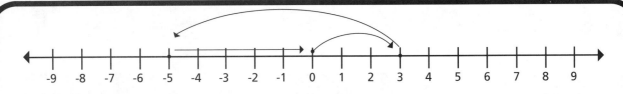

Hattie is recording the movements of a mouse across a grid. She notes that from the center, it runs 3" to the right, 8" to the left, and then returns to center.

$$3" + {}^-8" + 5" = 0$$

Write an equation and solve.

5) Cara monitors a snail in her aquarium. She notes that this morning it crawled 1" up the glass. A few hours later it crawled another 2" up the glass. Later, it crawled 4" down the glass. How far is the snail from where it started?

6) Henry starts the day with $30. He and his sister go to the fair. He pays for his own admission of $10, but he fails to notice that $15 falls out of his wallet while he is paying. His sister lends him $10 so that he can still enjoy the fair. Write the equation that illustrates his day's financial events.

7) Karen opens a credit account to purchase a new bed. She makes a down payment of $200 on a $1,000 bed. How much does she still owe the store?

Standards–Based Math • 5–6 © 2004 Creative Teaching Press

Subtracting Integers

Number and Operations

Subtract an integer by adding its opposite.
$^-8 - 5 =$
$^-8 + (^-5) = ^-13$

The negative of a negative is a positive.
$^-(^-a) = a$ $^-7 - (^-3) =$
            $^-7 + 3 = ^-4$

Rewrite each as an addition problem. Solve.

1 $^-72 - 70 = ^-72 +$ ____ = ____

2 $52 - (^-46) =$

3 $90 - 49 =$

4 $7 - (^-95) =$

5 $^-55 - (^-59) =$

6 $^-47 - 20 =$

Solve.

7 $57 + (^-70) =$

8 $^-32 - (^-88) =$

9 $^-84 + (^-61) =$

10 $^-18 - 39 =$

11 $21 + (^-24) =$

12 $79 - 55 =$

13 $97 + (^-99) =$

14 $68 - (^-73) =$

15 $^-92 + 43 =$

16 $49 + (^-19) =$

Multiplying Integers

NUMBER AND OPERATIONS

The **product** of a positive integer and a negative integer is a negative integer.
(a) (⁻b) = (⁻c)
The **product** of two negative integers OR two positive integers is a positive integer.
(a) (b) = (c)   (⁻a)(⁻b) = (c)

Solve.

1

Integers	=
(7)(⁻3)	
(⁻7)(3)	
(⁻7)(⁻3)	
(7)(3)	

2

Integers	=
(6)(⁻5)	
(⁻6)(5)	
(⁻6)(⁻5)	
(6)(5)	

3

Integers	=
(3)(⁻9)	
(⁻3)(9)	
(⁻3)(⁻9)	
(3)(9)	

4

Integers	=
(3)(⁻6)	
(⁻8)(9)	
(⁻5)(⁻7)	
(4)(3)	

5

Integers	=
(⁻7)(⁻2)	
(⁻8)(4)	
(3)(⁻8)	
(4)(9)	

6

Integers	=
(5)(8)	
(⁻8)(7)	
(⁻6)(⁻9)	
(5)(⁻3)	

7

Integers	=
(⁻5)(⁻8)(⁻2)	
(⁻2)(⁻6)(8)	
(3)(⁻7)(⁻4)	
(⁻3)(6)(⁻4)	

8

Integers	=
(2)(7)(⁻6)	
(⁻4)(5)(9)	
(⁻4)(⁻2)(⁻9)	
(⁻3)(⁻2)(6)	

9

Integers	=
(⁻3)(7)(7)	
(⁻5)(⁻4)(⁻5)	
(5)(⁻4)(⁻4)	
(⁻4)(7)(⁻9)	

Standards–Based Math • 5–6 © 2004 Creative Teaching Press

Dividing Integers

NUMBER AND OPERATIONS

> The **quotient** of a positive integer and a negative integer is a negative integer.
> a ÷ ⁻b = ⁻c
> The **quotient** of two negative integers OR two positive integers is a positive integer.
> a ÷ b = c   ⁻a ÷ ⁻b = c

Solve.

1

Integers	=
21 ÷ ⁻3	
⁻21 ÷ 3	
⁻21 ÷ ⁻3	
21 ÷ 3	

2

Integers	=
24 ÷ ⁻12	
⁻24 ÷ 12	
⁻24 ÷ ⁻12	
24 ÷ 12	

3

Integers	=
36 ÷ ⁻3	
⁻36 ÷ 3	
⁻36 ÷ ⁻3	
36 ÷ 3	

4

Integers	=
56 ÷ ⁻7	
⁻35 ÷ ⁻7	
⁻21 ÷ 7	
63 ÷ 7	

5

Integers	=
36 ÷ ⁻9	
⁻54 ÷ ⁻9	
⁻72 ÷ ⁻9	
99 ÷ ⁻9	

6

Integers	=
36 ÷ 12	
⁻60 ÷ 12	
84 ÷ ⁻12	
144 ÷ ⁻12	

7

Integers	=
400 ÷ ⁻80	
⁻52 ÷ ⁻4	
84 ÷ 6	
⁻108 ÷ 3	

8

Integers	=
⁻80 ÷ ⁻5	
175 ÷ 7	
⁻120 ÷ 8	
147 ÷ ⁻7	

9

Integers	=
⁻124 ÷ ⁻4	
⁻120 ÷ 24	
252 ÷ 6	
204 ÷ ⁻4	

Multiplying and Dividing Integers

NUMBER AND OPERATIONS

When both integers are positive or both are negative, the product or quotient is positive. When one
integer is positive and the other is negative, the
product or quotient is negative.

$$(a)(^-b) = (^-c)$$
$$(a)(b) = (c) \quad (^-a)(^-b) = (c)$$

$$a \div ^-b = ^-c$$
$$a \div b = c \quad ^-a \div ^-b = c$$

Solve.

1

Integers	=
(14)(⁻12)	
⁻182 ÷ 14	
(⁻120)(5)	
110 ÷ 5	

2

Integers	=
(⁻6)(⁻86)	
⁻266 ÷ 7	
(12)(⁻15)	
140 ÷ 5	

3

Integers	=
(⁻7)(⁻53)	
(12)(16)	
⁻196 ÷ ⁻14	
132 ÷ 22	

4

Integers	=
120 ÷ ⁻5	
(⁻8)(⁻27)	
⁻376 ÷ 8	
(⁻19)(12)	

5

Integers	=
(⁻12)(17)	
⁻160 ÷ ⁻8	
⁻414 ÷ ⁻9	
(3)(⁻84)	

6

Integers	=
(⁻9)(44)	
⁻224 ÷ 16	
(7)(⁻29)	
192 ÷ ⁻6	

7

Integers	=
105 ÷ ⁻7	
(35)(7)	
(⁻7)(7)(⁻7)	
⁻312 ÷ 12	

8

Integers	=
⁻252 ÷ −14	
(⁻8)(⁻5)(⁻3)	
(15)(⁻2)(⁻4)	
600 ÷ ⁻8	

9

Integers	=
⁻224 ÷ ⁻32	
(12)(⁻7)(⁻3)	
144 ÷ 9	
(16)(⁻4)(4)	

Checks and Balances

NUMBER AND OPERATIONS

A check register is used to record transactions in a checking account. When you write a check, you are taking money from the account. This amount is written in the withdrawal column. When you deposit funds, you place money in the account. The balance column records the total positive or negative amount of money in the account.

Complete the balance column.

	Date	Item	Withdrawal −	Deposit +	Balance =
	9/1				$1,800.50
1.	9/1	Mortgage	$880.00		
2.	9/2	Groceries	$250.00		
3.	9/3	Cash	$80.00		
4.	9/5	Car Payment	$355.86		
5.	9/5	Phone Bill	$75.65		
6.	9/7	Gift		$40.00	
7.	9/8	Water Bill	$72.65		
8.	9/9	Electric Bill	$289.46		
9.	9/9	Pet Food	$35.23		
10.	9/10	Sale of Old Car		$750.00	
11.	9/15	Paycheck		$1,200.00	
12.	9/16	Student Loan	$250.00		
13.	9/16	Groceries	$265.98		
14.	9/17	Car Insurance	$365.00		

Mental Math with Zeros

NUMBER AND OPERATIONS

40 × 600 4 × 6 = 24 Count the zeros in the factors. Place the same number in the product. 40 × 600 = 24,000	24,000 ÷ 600 Cross out an equal number of zeros in both numbers. Divide the remaining numbers to find the quotient. 240 ÷ 6 = 40

Use mental math to solve.

1 70 × 400 =

2 3,600 ÷ 600 =

3 400 × 800 =

4 180,000 ÷ 6,000 =

5 60 × 30,000 =

6 14,000 ÷ 700 =

7 90 × 20 =

8 42,000 ÷ 6,000 =

9 8,000 × 600 =

10 560,000 ÷ 8,000 =

11 90 × 8,000 =

12 8,100 ÷ 900 =

13 500 × 700 =

14 28,000 ÷ 700 =

Standards–Based Math • 5–6 © 2004 Creative Teaching Press

Multiplying Larger Numbers

NUMBER AND OPERATIONS

```
        3  2  6   factor
    ×   5  9  2   factor
        6  5  2   1st partial product
     2  9  3  4   2nd partial product
  + 1  6  3  0    3rd partial product
  1  9  2  9  9  2   product
```

Solve.

1   273
      × 975

2   554
      × 328

3   933
      × 874

4   857
      × 342

5   716
      × 195

6   987
      × 225

7   619
      × 278

8   369
      × 751

9   915
      × 562

10   693
       × 824

11   801
       × 703

12   147
       × 351

Name _____ Date _____

Fraction Review

NUMBER AND OPERATIONS

The **numerator** names a specific portion of the whole.
The **denominator** names the total number of equal parts within the whole.

 $\frac{1}{8}$ $\frac{1}{3}$ $\frac{3}{4}$

Write the fraction.

1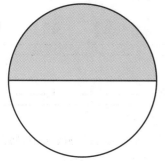

_____

2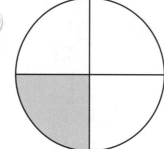

_____

3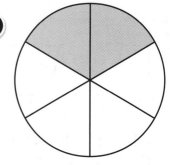

_____

4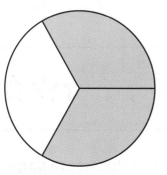

_____

5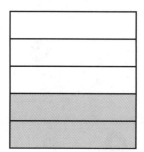

_____

6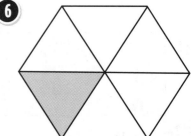

_____

7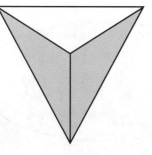

_____

8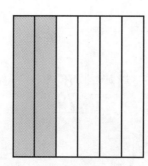

_____

9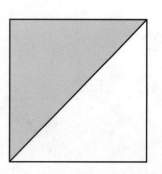

_____

Standards–Based Math • 5–6 © 2004 Creative Teaching Press

Fraction of a Set

Number and Operations

> The **numerator** names a specific portion of the whole.
> The **denominator** names the total number of equal parts within the whole.
> A whole can be a set of objects.
>
> $\frac{3}{4}$ of the set are cylinders.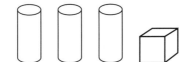

Write the fraction for each set.

Which portion of the set is shaded?
Which portion is not shaded?

 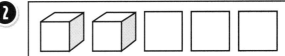

Which portion of the set are cubes?
Which portion are squares?

Which portion of the set are donuts?
Which portion is a solid circle?

Which portion of the set are large
stars? Which portion are small stars?

 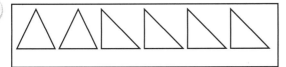

Which portion of the set are isosceles
triangles? Which portion are right
triangles?

 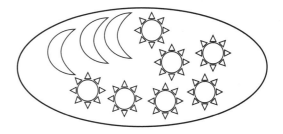

Which portion of the set are moons?
Which portion are suns?

Name _____ Date _____

Converting Fractions to Decimals

NUMBER AND OPERATIONS

Convert $\frac{1}{2}$ to a decimal.

Divide the numerator by the denominator. $2\overline{)1.0}$ with 0.5 above

Convert ea~~ch~~

9/19

1 $\dfrac{1}{5}$

2 $\dfrac{3}{12}$

3 $\dfrac{15}{50}$

4 $\dfrac{24}{60}$

5 $\dfrac{1}{20}$

6 $\dfrac{2}{40}$

7 $\dfrac{4}{5}$

8 $\dfrac{24}{30}$

9 $\dfrac{1}{10}$

10 $\dfrac{4}{4}$

11 $\dfrac{5}{8}$

12 $\dfrac{2}{5}$

13 $\dfrac{3}{20}$

14 $\dfrac{6}{75}$

Standards-Based Math • 5-6 © 2004 Creative Teaching Press

Name _____ Date _____

Converting Fractions to Repeating Decimals

Number and Operations

Convert $\frac{1}{3}$ to a decimal.
Divide the numerator by the denominator.

$$
\begin{array}{r}
0.\overline{333} \\
3\overline{)1.000} \\
-09 \\
\hline
010 \\
-009 \\
\hline
0010 \\
-0009 \\
\hline
001
\end{array}
$$

A line over the last digit(s) of a decimal means that the decimal repeats those numbers from that point on.

Convert each fraction to a repeating decimal.

1 $\dfrac{2}{6}$

2 $\dfrac{5}{18}$

3 $\dfrac{4}{12}$

4 $\dfrac{5}{9}$

5 $\dfrac{4}{21}$

6 $\dfrac{5}{24}$

7 $\dfrac{7}{27}$

8 $\dfrac{4}{33}$

9 $\dfrac{8}{24}$

10 $\dfrac{5}{54}$

Converting Decimals to Percents

NUMBER AND OPERATIONS

Convert a fraction that simplifies to a decimal to a percent by multiplying by 100.

$\frac{1}{2} = 0.5$

$0.5 \times 100 = 50\%$

$\frac{1}{2} = 0.5 = 50\%$

Convert each decimal to a percent.

1 $\quad \frac{1}{5} = 0.2 =$

2 $\quad \frac{3}{12} = 0.25 =$

3 $\quad \frac{15}{50} = 0.3 =$

4 $\quad \frac{24}{60} = 0.4 =$

5 $\quad \frac{1}{20} = 0.05 =$

6 $\quad \frac{2}{40} = 0.05 =$

7 $\quad \frac{4}{5} = 0.8 =$

8 $\quad \frac{24}{30} = 0.8 =$

9 $\quad \frac{1}{10} = 0.1 =$

10 $\quad \frac{4}{4} = 1 =$

11 $\quad \frac{5}{8} = 0.625 =$

12 $\quad \frac{2}{5} = 0.4 =$

13 $\quad \frac{3}{20} = 0.15 =$

14 $\quad \frac{6}{75} = 0.08 =$

Approximate Percents

NUMBER AND OPERATIONS

Convert a fraction that simplifies to a repeating decimal to a percent by rounding the decimal to the hundreths place. Then multiply by 100.

$$\frac{1}{6} = 0.16\bar{6}$$
$$0.166 \approx 0.17$$
$$0.17 \times 100 = 17$$
So, $\frac{1}{6} = 0.166 \approx 17\%$

Solve.

1 $\dfrac{2}{6} = 0.333 \approx$ _____ $\times\ 100 =$

2 $\dfrac{5}{18} = 0.2777 \approx$ _____ $\times\ 100 =$

3 $\dfrac{4}{12} = 0.333 \approx$ _____ $\times\ 100 =$

4 $\dfrac{5}{9} = 0.555 \approx$ _____ $\times\ 100 =$

5 $\dfrac{4}{21} = 0.190476 \approx$ _____ $\times\ 100 =$

6 $\dfrac{5}{24} = 0.208333 \approx$ _____ $\times\ 100 =$

7 $\dfrac{7}{27} = 0.259259 \approx$ _____ $\times\ 100 =$

8 $\dfrac{4}{33} = 0.1212 \approx$ _____ $\times\ 100 =$

9 $\dfrac{8}{24} = 0.333 \approx$ _____ $\times\ 100 =$

10 $\dfrac{5}{54} = 0.0925925 \approx$ _____ $\times\ 100 =$

Multiples

Number and Operations

A **multiple** is the product of a given factor by another whole number.
6, 12, 18, 24, 30, and 36 are multiples of 6. They are also multiples of 3.

List the first 10 multiples of each factor.

1 7: _____

2 12: _____

3 3: _____

4 9: _____

5 4: _____

6 11: _____

7 8: _____

8 5: _____

9 10: _____

10 2: _____

List the first 10 multiples of each pair of factors. Circle any multiples they have in common.

11 6: _____

   9: _____

12 4: _____

   12: _____

Standards–Based Math • 5–6 © 2004 Creative Teaching Press

Name _____ Date _____

Least Common Multiple

NUMBER AND OPERATIONS

The **least common multiple** is the multiple a pair or group of numbers have in common with the least value.
4: 4, 8, 12, 16, 20, **24,** 28, 32, 36, 40, 44, 48, 52, 56, 60
6: 6, 12, 18, **24,** 30, 36, 42, 48, 54, 60
8: 8, 16, **24,** 32, 40, 48, 56, 64
There is only one least common multiple for any group of numbers.
24 is the least common multiple of 4, 6, and 8.

List multiples of each number. Circle the least common multiple for each group of numbers.

1 5: _____
   8: _____

2 7: _____
   5: _____

3 3: _____
   8: _____

4 9: _____
   12: _____
   6: _____

5 15: _____
   10: _____
   20: _____

6 12: _____
   16: _____

Greatest Common Factor

NUMBER AND OPERATIONS

The **greatest common factor** is the factor a pair or group of numbers have in common with the greatest value.
4: 4, **2,** 1   **36:** 36, 18, 9, **12,** 6, 4, 3, 2, 1
6: 6, 3, **2,** 1   **48:** 48, 24, 19, **12,** 8, 6, 4, 3, 2, 1
8: 8, 4, **2,** 1
There is only one greatest common factor for any group of numbers.

List the factors. Circle the greatest common factor for each group of numbers.

1 24: _____

   36: _____

2 36: _____

   9: _____

3 21: _____

   28: _____

4 32: _____

   40: _____

   16: _____

5 45: _____

   15: _____

   60: _____

6 12: _____

   16: _____

Name _____ Date _____

Reducing Fractions

Number and Operations

> To convert a fraction to its simplest terms, divide the numerator and denominator by their greatest common factor.
>
> $$\frac{21}{28} \div \frac{7}{7} = \frac{3}{4}$$

Reduce to simplest terms.

1 $\dfrac{6}{63} =$

2 $\dfrac{36}{60} =$

3 $\dfrac{8}{20} =$

4 $\dfrac{18}{24} =$

5 $\dfrac{24}{32} =$

6 $\dfrac{12}{16} =$

7 $\dfrac{20}{60} =$

8 $\dfrac{32}{48} =$

9 $\dfrac{9}{54} =$

10 $\dfrac{10}{16} =$

11 $\dfrac{50}{75} =$

12 $\dfrac{18}{32} =$

13 $\dfrac{31}{93} =$

14 $\dfrac{11}{88} =$

15 $\dfrac{32}{96} =$

16 $\dfrac{7}{35} =$

Adding Fractions

NUMBER AND OPERATIONS

Use the least common multiple to put fractions in the same terms. Add numerators. Reduce to simplest terms using the greatest common factor.

$$\frac{1}{3} + \frac{1}{6} = \frac{2}{6} + \frac{1}{6} = \frac{3}{6} = \frac{1}{2}$$

Add and reduce to simplest terms.

1 $\dfrac{2}{3} + \dfrac{1}{4} =$

2 $\dfrac{2}{11} + \dfrac{7}{11} =$

3 $\dfrac{10}{15} + \dfrac{2}{15} =$

4 $\dfrac{1}{6} + \dfrac{3}{4} =$

5 $\dfrac{6}{11} + \dfrac{3}{22} =$

6 $\dfrac{13}{24} + \dfrac{5}{12} =$

7 $\dfrac{4}{15} + \dfrac{3}{5} =$

8 $\dfrac{7}{10} + \dfrac{3}{15} =$

9 $\dfrac{9}{12} + \dfrac{1}{8} =$

10 $\dfrac{3}{8} + \dfrac{1}{24} =$

11 $\dfrac{2}{3} + \dfrac{1}{6} + \dfrac{1}{12} =$

12 $\dfrac{1}{2} + \dfrac{3}{8} + \dfrac{1}{12} =$

Name _____ Date _____

Subtracting Fractions

NUMBER AND OPERATIONS

Use the least common multiple to put fractions in the same terms. Subtract numerators. Reduce to simplest terms using the greatest common factor.

$$\frac{2}{3} - \frac{1}{6} = \frac{4}{6} - \frac{1}{6} = \frac{3}{6} = \frac{1}{2}$$

Subtract and reduce to simplest terms.

1 $\frac{2}{3} - \frac{2}{4} =$

2 $\frac{7}{11} - \frac{2}{11} =$

3 $\frac{12}{15} - \frac{2}{15} =$

4 $\frac{5}{6} - \frac{3}{4} =$

5 $\frac{6}{11} - \frac{4}{22} =$

6 $\frac{9}{12} - \frac{13}{24} =$

7 $\frac{3}{5} - \frac{4}{15} =$

8 $\frac{7}{10} - \frac{3}{15} =$

9 $\frac{10}{12} - \frac{9}{24} =$

10 $\frac{3}{8} - \frac{1}{24} =$

11 $\frac{2}{15} - \frac{1}{30} - \frac{1}{45} =$

12 $\frac{1}{2} - \frac{1}{3} - \frac{1}{12} =$

Improper Fractions to Mixed Numbers

Number and Operations

Divide the numerator by the denominator. Place a remainder over the denominator. Simplify.

$$\frac{26}{8} \longrightarrow 8\overline{)26}^{\,3\,R2} \longrightarrow \frac{26}{8} = 3\frac{2}{8} = 3\frac{1}{4}$$

Simplify.

1 $\dfrac{10}{6}$

2 $\dfrac{30}{8}$

3 $\dfrac{28}{9}$

4 $\dfrac{17}{7}$

5 $\dfrac{13}{2}$

6 $\dfrac{34}{5}$

7 $\dfrac{39}{11}$

8 $\dfrac{14}{3}$

9 $\dfrac{68}{12}$

10 $\dfrac{30}{19}$

11 $\dfrac{127}{19}$

12 $\dfrac{23}{4}$

13 $\dfrac{44}{10}$

14 $\dfrac{13}{6}$

15 $\dfrac{48}{10}$

16 $\dfrac{49}{11}$

17 $\dfrac{29}{8}$

18 $\dfrac{31}{6}$

Standards-Based Math • 5-6 © 2004 Creative Teaching Press

Mixed Numbers to Improper Fractions

Number and Operations

$4\dfrac{3}{5}$

Step 1: Multiply the denominator by the whole number. $5 \times 4 = 20$
Step 2: Add the product to the numerator. $20 + 3 = 23$
Step 3: Place the sum over the denominator. $\dfrac{23}{5}$

$4\dfrac{3}{5} = \dfrac{23}{5}$

Convert to an improper fraction.

1 $7\dfrac{5}{7} =$

2 $4\dfrac{1}{11} =$

3 $11\dfrac{2}{5} =$

4 $8\dfrac{2}{9} =$

5 $6\dfrac{3}{10} =$

6 $8\dfrac{2}{8} =$

7 $12\dfrac{5}{6} =$

8 $12\dfrac{3}{6} =$

9 $8\dfrac{6}{9} =$

10 $9\dfrac{1}{8} =$

11 $6\dfrac{4}{12} =$

12 $12\dfrac{4}{5} =$

13 $8\dfrac{3}{8} =$

14 $9\dfrac{1}{6} =$

15 $9\dfrac{1}{5} =$

16 $11\dfrac{1}{5} =$

17 $7\dfrac{2}{11} =$

18 $10\dfrac{5}{6} =$

Name _____ Date _____

Multiplying Fractions

NUMBER AND OPERATIONS

$\dfrac{2}{10} \times \dfrac{6}{9} =$

Step 1: Simplify fractions, if needed. $\dfrac{1}{5} \times \dfrac{2}{3} =$

Step 2: Multiply the numerators.
Multiply the denominators. $\dfrac{1 \times 2}{5 \times 3} = \dfrac{2}{15}$

Step 3: Simplify again, if needed. $\dfrac{2}{15}$

$\dfrac{2}{10} \times \dfrac{6}{9} = \dfrac{2}{15}$

Multiply.

1 $\dfrac{3}{9} \times \dfrac{3}{7} =$

2 $\dfrac{3}{4} \times \dfrac{7}{8} =$

3 $\dfrac{2}{5} \times \dfrac{1}{3} =$

4 $\dfrac{6}{8} \times \dfrac{1}{3} =$

5 $\dfrac{4}{6} \times \dfrac{5}{9} =$

6 $\dfrac{5}{7} \times \dfrac{1}{5} =$

7 $\dfrac{3}{4} \times \dfrac{2}{4} =$

8 $\dfrac{4}{7} \times \dfrac{2}{4} =$

9 $\dfrac{1}{3} \times \dfrac{4}{7} =$

10 $\dfrac{8}{9} \times \dfrac{4}{8} =$

11 $\dfrac{1}{3} \times \dfrac{6}{7} =$

12 $\dfrac{3}{4} \times \dfrac{1}{5} =$

13 $\dfrac{1}{7} \times \dfrac{5}{7} =$

14 $\dfrac{3}{5} \times \dfrac{1}{5} =$

15 $\dfrac{5}{8} \times \dfrac{3}{7} =$

16 $\dfrac{4}{8} \times \dfrac{1}{7} =$

Multiplying Mixed Numbers

NUMBER AND OPERATIONS

Change to improper fractions. Multiply numerators and denominators. Simplify.

$$1\frac{2}{9} \times 3\frac{3}{4} = \frac{11}{9} \times \frac{15}{4} = \frac{11 \times 15}{9 \times 4} = \frac{165}{36} = 4\frac{21}{36} = 4\frac{7}{12}$$

Multiply.

1 $\quad 4\frac{1}{3} \times 5\frac{3}{8} =$

2 $\quad 2\frac{2}{3} \times 4\frac{3}{4} =$

3 $\quad 3\frac{1}{3} \times 4\frac{5}{8} =$

4 $\quad 5\frac{3}{8} \times 4\frac{1}{4} =$

5 $\quad 3\frac{7}{8} \times 1\frac{7}{9} =$

6 $\quad 4\frac{2}{3} \times 5\frac{5}{8} =$

7 $\quad 9\frac{1}{3} \times 3\frac{4}{5} =$

8 $\quad 4\frac{4}{7} \times 1\frac{15}{16} =$

Name _____ Date _____

Dividing Fractions

NUMBER AND OPERATIONS

Multiply the dividend by the reciprocal of the divisor. Simplify.

$$\frac{3}{4} \div \frac{1}{2} = \frac{3 \times 2}{4 \times 1} = \frac{6}{4} = 1\frac{1}{2}$$

Divide.

1 $\frac{7}{8} \div \frac{2}{6} =$

2 $\frac{1}{9} \div \frac{2}{4} =$

3 $\frac{4}{7} \div \frac{4}{5} =$

4 $\frac{1}{3} \div \frac{2}{5} =$

5 $\frac{3}{5} \div \frac{7}{9} =$

6 $\frac{3}{8} \div \frac{4}{7} =$

7 $\frac{6}{9} \div \frac{2}{5} =$

8 $\frac{4}{8} \div \frac{5}{7} =$

9 $\frac{2}{4} \div \frac{5}{6} =$

10 $\frac{4}{7} \div \frac{3}{4} =$

11 $\frac{4}{6} \div \frac{2}{5} =$

12 $\frac{2}{8} \div \frac{1}{3} =$

13 $\frac{3}{7} \div \frac{3}{9} =$

14 $\frac{2}{8} \div \frac{6}{7} =$

15 $\frac{5}{6} \div \frac{6}{5} =$

Standards-Based Math • 5–6 © 2004 Creative Teaching Press

Decimals to Fractions

NUMBER AND OPERATIONS

Convert the decimal to a fraction with 1 as the denominator. Multiply the numerator and denominator by 1, expressed as a fraction with the numerator and denominator each equal to the place of the last digit of the decimal. Simplify.

$$2.5 = \frac{2.5}{1} \times 1 = \frac{2.5}{1} \times \frac{10}{10} = \frac{25}{10} = \frac{5}{2} = 2\frac{1}{2}$$

$$2.35 = \frac{2.35}{1} \times 1 = \frac{2.35}{1} \times \frac{100}{100} = \frac{235}{100} = \frac{47}{20} = 2\frac{7}{20}$$

Convert to a fraction.

1 4.25

2 0.48

3 3.7

4 0.004

5 4.132

6 0.09

7 1.75

8 0.25

9 0.4

10 2.8

Percent of a Set

NUMBER AND OPERATIONS

 $\frac{1}{3} = 0.\overline{33} \approx 33\%$

Write the percent that is shaded. Estimate if needed.

1 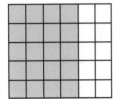 _____

2   _____

3  _____

4 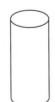 _____

5  _____

6 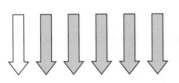 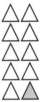 _____

Shade the given portion of the set.

7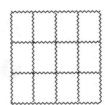

33%

8 _____

25%

9

66%

10

75%

11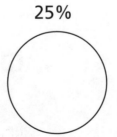

50%

12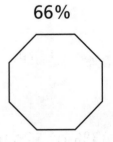

25%

Percent of a Value

NUMBER AND OPERATIONS

Change the percent to a decimal. Then multiply the given value by that decimal.

4% of 20 =

$$
\begin{array}{r}
2\ 0 \\
\times\ 0.\ 0\ 4 \\
\hline
0.\ 8\ 0
\end{array}
$$

4% of 20 = 0.80

Calculate the value of each percent.

1 33% of $12.00

2 98% of 250

3 120% of 625

4 9% of 60

5 20% of 80

6 2% of 1200 mL

7 82% of 6 m

8 25% of $550.00

9 6% of 380

10 32% of 297

11 25% of $60.00

12 90% of 120

13 50% of 200

14 25% of 80

15 33% of $24.00

16 55% of 28

Name _____   Date _____

Leaving a Tip

NUMBER AND OPERATIONS

A standard tip is 15% percent of the bill. To calculate an accurate tip mentally, find 10% and round to the hundredths place. Find half of that number (5% of total). Add to 10% to find a 15% tip.

$32.58
$3.258 = $3.26 $3.26 ÷ 2 = $1.63
15% = $1.63 + $3.26 = $4.89

Calculate a 15% tip using mental math.

1 $6.50

2 $45.00

3 $22.85

4 $18.40

5 $7.85

6 $14.22

A generous tip to show appreciation for good service is 20% of the bill. Find 10% and round to the hundredths place. Double to find a 20% tip.

$32.58
$3.258 = $3.26   20% = $6.52

Calculate a 20% tip using mental math.

7 $40.00

8 $35.72

9 $21.98

10 $7.65

Standards–Based Math • 5–6 © 2004 Creative Teaching Press

Name _____   Date _____

Paying Sales Tax

NUMBER AND OPERATIONS

Determine sales tax by multiplying the percent of tax by the total item cost.

7.5% tax on clothes

$$
\begin{array}{r}
\$\ 1\ 4.\ 0\ 0\ 0 \\
\times\ \ 0.\ 0\ 7\ 5 \\
\hline
7\ 0\ 0\ 0\ 0 \\
1\ 9\ 8\ 0 \\
\hline
\$\ 1.\ 0\ 5
\end{array}
$$

$1.05 tax
$15.05 total

5.0% gasoline tax

$$
\begin{array}{r}
\$\ 1\ 4.\ 0\ 0 \\
\times\ \ 0.\ 0\ 5 \\
\hline
0.\ 7\ 0\ 0\ 0
\end{array}
$$

70¢ tax
$14.70 total

Calculate the sales tax.

1 $15.72 if tax is 7%

2 $7.75 if tax is 6%

3 $12.95 if tax is 6.5%

4 $15.82 if tax is 5%

5 $15.82 if tax is 8%

6 $85.00 if tax is 12%

7 $67.29 if tax is 7.5%

8 $32.33 if tax is 7%

Determine the total cost without adding by multiplying 1 + the percent of sales tax.

5.0% gasoline tax

$$
\begin{array}{r}
\$\ 1\ 4.\ 0\ 0 \\
\times\ \ 1.\ 0\ 5 \\
\hline
0.\ 7\ 0\ 0\ 0 \\
+\ 1\ 4.\ 0\ 0 \\
\hline
\$\ 1\ 4.\ 7\ 0
\end{array}
$$

Calculate the total with sales tax.

9 $14.99 if tax is 6%

10 $15.72 if tax is 6.5%

11 $12.95 if tax is 12.5%

12 $15.82 if tax is 5%

13 $67.29 if tax is 7.5%

14 $32.33 if tax is 8%

Name _____ Date _____

Analogies 1

ALGEBRA

◯ : ◯ as ▢ : ▢

: means "is to"

Determine the relationship ~~9/19~~ of the choice that completes the analogy.

1 : 3   as   ▢ : _____   **a.** ▢

2  as   ▢ : _____   **b.**

3 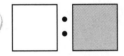 as   ◯ : _____   **c.** **4**

4  as : _____   **d.**

5  as : _____   **e.**

6  as : _____   **f.**

7 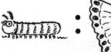 as : _____   **g.**

Standards–Based Math • 5–6 © 2004 Creative Teaching Press

Analogies 2

Algebra

Analyze the relationship in the first set and apply the same "rule" to the second set. There may be more than one right answer. Choose the best answer.

$4^2 : 16$ $5^2 :$ __
a. 10
b. 25
c. 15

Complete each analogy by writing the letter of the best choice on the line.

1 $36 : 6^2$ as $64 :$ ____

     a. 8

     b. 8^2

     c. (6)(8)

2 $4 : -4$ as $-9 :$ ____

     a. 9

     b. -9

     c. -1

3 $9 : 54$ as $6 :$ ____

     a. 32

     b. 9

     c. 36

4 $9 : 18$ as $6 :$ ____

     a. 54

     b. 18

     c. 15

5 $45 : 450$ as $345 :$ ____

     a. 3,450

     b. 34.5

     c. 3.45

6 $25\% : \dfrac{1}{4}$ as $62.5\% :$ ____

     a. $\dfrac{2}{3}$

     b. $\dfrac{5}{8}$

     c. $\dfrac{5}{9}$

Patterns

Algebra

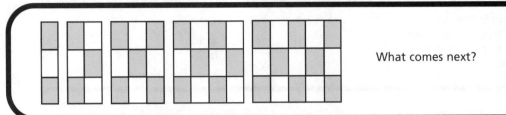 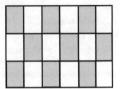

What comes next?

Draw the next three images in the pattern.

1

2

3

4

5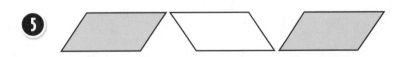

6

Standards–Based Math • 5–6 © 2004 Creative Teaching Press

Number Patterns 1

Algebra

8 goes in, 32 comes out.
4 goes in, 16 comes out.
What is the rule?
× 4

Determine the rule.

1 4 goes in, 20 comes out.
6 goes in, 30 comes out.

2 14 goes in, 28 comes out.
9 goes in, 23 comes out.

3 12 goes in, 4 comes out.
39 goes in, 13 comes out.

4 12 goes in, 36 comes out.
9 goes in, 27 comes out.

5 31 goes in, 24 comes out.
65 goes in, 58 comes out.

6 42 goes in, 7 comes out.
84 goes in, 14 comes out.

7 9 goes in, 108 comes out.
12 goes in, 144 comes out.

8 16 goes in, 14 comes out.
87 goes in, 85 comes out.

9 76 goes in, 84 comes out.
25 goes in, 33 comes out.

10 14 goes in, 84 comes out.
20 goes in, 120 comes out.

11 45 goes in, 76 comes out.
26 goes in, 57 comes out.

12 112 goes in, 14 comes out.
200 goes in, 25 comes out.

Name _____ Date _____

Number Patterns 2

ALGEBRA

Evaluate number patterns by looking at the difference between each number and then looking at the numbers as a whole. Some patterns repeat and some do not.

3, 6, 9, 12, 15, 18
pattern: +3

Continue the pattern.

1 3, 5, 7, 9, 11, _____, _____, _____

2 95, 90, 85, 80, 75, 70, _____, _____, _____

3 4, 9, 7, 12, 10, 15, 13, 18, 16, _____, _____, _____

4 18, 36, 54, 72, 90, _____, _____, _____

5 21, 28, 35, 42, _____, _____, _____

6 10, 20, 30, 40, 50, _____, _____, _____

7 5, 10, 20, 35, 55, _____, _____, _____

8 5, 12, 19, 26, _____, _____, _____

9 1, 10, 19, 28, _____, _____, _____

10 1, 2, 3, 5, 7, 11, 13, 17, 19, 23, _____, _____, _____

Standards–Based Math • 5–6 © 2004 Creative Teaching Press

Order of Operations

Algebra

Solve equations according to the rules outlined in the order of operations. First, simplify the expressions in parentheses. Save addition and subtraction for last.

Parentheses
Exponents
Multiplication
Division
Addition
Subtraction

Solve.

1 $7 + (8 - 2) \times 2$

2 $\dfrac{4^2}{2^2} =$

3 $(5 + 3)^2$

4 $(4 \times 2) - 3$

5 $\dfrac{(8 + 2)^2}{(4 + 1)^2} + 4 =$

6 $(50 + 6) \div 7 =$

7 $2(3 + 5)^2 - 3$

8 $31 + (3)(10)$

9 $[2(2 + 4) + 4] \times 3 + 2$

10 $5 + \dfrac{9 + 5}{3 + 4}$

11 $\dfrac{1}{2}(4 + 4)$

12 $(3 + 4)(7 + 1)$

13 $\left(\dfrac{2^3}{16}\right)(6)$

14 $\dfrac{1}{3}(6 + 4 + 2) + 8$

Commutative Property

ALGEBRA

The **commutative property** says you can switch the order of two numbers and still get the same answer.

2 + 8 = 8 + 2

(5)(3) = (3)(5)

Solve each problem in the table. Then write two equations for the problem on the lines.

×	2	4	$\frac{1}{2}$	-3	5
$\frac{1}{3}$					
6					
$\frac{3}{5}$					
5					
8					

$2 \times \frac{1}{3}$; $\frac{1}{3} \times 2$

_____

_____

_____

_____

_____

_____

_____

_____

_____

_____

_____

_____

_____

_____

Standards–Based Math • 5–6 © 2004 Creative Teaching Press

Associative Property

ALGEBRA

> The **associative property** says the way three numbers are grouped for only addition or multiplication will not affect the outcome.
>
> $$2 + (4 + 3) \qquad 2\,(4 \times 6)$$
> $$(2 + 4) + 3 \qquad (2 \times 4)\,6$$

Rewrite each problem to demonstrate the associative property. Solve.

1 $(2 \times 3) \times 8$

2 $12 + (4 + 3)$

3 $(5 \times 4) \times 3$

4 $8 + (12 + 3)$

5 $(20 + 4) + 18$

6 $(4 \times 3) \times 5$

7 $2 \times (3 \times 7)$

8 $25 + (48 + 36)$

9 $(21 + 14) + 9$

10 $4 \times (3 \times 5)$

11 $203 + (404 + 362)$

12 $3 \times (4 \times 4)$

13 $(298 + 45) + 36$

14 $(2 \times 9) \times 6$

Distributive Property

ALGEBRA

The **distributive property** says that when a number is multiplied by the sum or difference of two other numbers, the first number can be distributed to both of those two numbers and multiplied by each of them separately.

$$a(b + c) = ab + ac$$
$$5(4 + 3) = 5 \times 4 + 5 \times 3$$
$$5(7) = 20 + 15$$
$$35 = 35$$

Rewrite each equation to distribute the multiplication. Solve both sides to check your work.

1 $3(5 + 2) =$

2 $^-4(5 + 3) =$

3 $2(5 + 6) =$

4 $^-(4 - 2)5 =$

5 $7(6 + 2) =$

6 $5(18 - 9) =$

7 $(2 + {}^-6)8 =$

8 $(5 + 6)3 =$

9 $7(7 - 3) =$

Exponents

A L G E B R A

An **exponent** next to a number is a short way of writing a multiplication problem in which a number is multiplied by itself. The exponent tells how many times the number appears in the multiplication problem.

$2 \times 2 = 2^2$

$2 \times 2 \times 2 \times 2 = 2^4$

We read a number with an exponent as (number) to the (exponent) power.

2^4 = two to the fourth power 2^2 = two to the second power

Write out the multiplication problem represented by each number and exponent. Solve.

1 5^3

2 2^4

3 3^4

4 6^2

5 8^3

6 9^2

7 5^4

8 7^3

9 4^3

10 10^3

11 12^2

12 11^3

Standards-Based Math • 5–6 © 2004 Creative Teaching Press

Name _____ Date _____

Exponential Notation

ALGEBRA

> The number 10 with an exponent is special. The exponent tells you exactly how many zeros come after the 1.
>
> $$10^6 = 10 \cdot 10 \cdot 10 \cdot 10 \cdot 10 \cdot 10 = 1,000,000$$

Write out the number each power of ten represents.

1 10^3

2 10^8

3 10^7

> Scientific notation uses powers of ten to write long numbers in a short way. The power of ten tells you how many places to move the decimal point and in which direction. A **positive exponent** means move the decimal point to the right. A **negative exponent** means move the decimal point to the left.
>
> $$8,340,000,000,000 = 8.34 \times 10^{12}$$
> $$0.00000000005 = 5 \times 10^{-11}$$

Write out the number each power of ten represents.

4 7.2×10^5

5 6×10^7

6 8.7×10^{-4}

7 9×10^9

8 2×10^{-8}

9 1.204×10^{-6}

10 6×10^{-7}

11 2×10^{12}

12 1.11×10^{10}

Standards–Based Math • 5–6 © 2004 Creative Teaching Press

Variables

ALGEBRA

A **variable** is a letter or symbol that stands for a number.

$$7 + n \longleftarrow \text{variable}$$

Variables can be helpful when you want to solve a word problem.

Charles bought 5 new cards for a total of twenty-five cards in his collection. How many cards did he have before the purchase? You could draw a model:

$$?? + \square = \square$$

But it is faster to write an equation with a variable.

Original cards $\longrightarrow$ $n + 5 = 25$

Draw a model. Then write an equation. Use variable n.

1 Henry has three books in the *Up and 'Attem Boys* series. He buys a few more to give him a total of eight books in the series. How many books did he buy?

2 Kylie drove 200 miles today for a total of 1,000 miles traveled on this trip. How many miles had she driven on the trip before today?

Write an equation for the problem. Use variable x.

3 Max and Ben invite some friends over to play. Now there are 7 children playing at the house. How many friends came over?

4 Cora pulled another batch of cookies out of the oven. Placing the 12 cookies on the rack to cool, she said, "There! That's 48 cookies altogether!" How many cookies were done before this batch?

Name _____   Date _____

Inverse Operations

ALGEBRA

If you move three whole numbers to the right on the number line (+3), how do you get back to where you started?

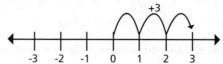

You do the inverse operation with the same value (–3).

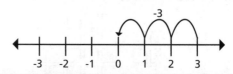

Match each operation and value to its inverse.

_____   **1** +5   **a.** ÷ (–4)

_____   **2** –10   **b.** + 1.07

_____   **3** × 3   **c.** + 4

_____   **4** ÷ 6   **d.** – 143

_____   **5** + (–4)   **e.** – 5

_____   **6** ÷ 0.5   **f.** × 6

_____   **7** – 1.07   **g.** + 10

_____   **8** + 143   **h.** – 19

_____   **9** + $\dfrac{6}{2}$ **i.** × 8

_____   **10** ÷ 8   **j.** ÷ 3

_____   **11** × (–4)   **k.** × 0.5

_____   **12** + 19   **l.** – 3

Standards–Based Math • 5–6 © 2004 Creative Teaching Press

Inverse Operations: Addition

ALGEBRA

> Use subtraction to isolate the variable when addition is indicated.
> Subtract the value from both sides.
>
> $22 = x + 6$
> $22 - 6 = x + 6 - 6$
> $16 = x$

Solve.

1 $x + 4 = 12$

2 $a + 6 = 24 + 9$

3 $y + 5 = 18$

4 $z + (5 + 3) = 50$

5 $x + (4 - 2) = 12$

6 $a + (3 \times 5) = 100$

7 $15 + x = 100$

8 $9 + a = 27$

9 $(7 + 9) + y = 75$

10 $(6 \times 4) + b = 60$

11 $(9 - 4) + x = 25$

12 $8 + c = 8$

Name _____   Date _____

Inverse Operations: Subtraction

ALGEBRA

Use addition to isolate the variable when subtraction is indicated.
Add the value to both sides.

$32 = x - 2$
$32 + 2 = x - 2 + 2$
$34 = x$

Solve.

1 $x - 4 = 12$

2 $a - 6 = 15 + 9$

3 $y - 5 = 20$

4 $z - (7 + 3) = 85$

5 $x - (9 - 2) = 78$

6 $a - (9 \times 5) = 100$

7 $^{-}15 + x = 100$

8 $^{-}6 + a = 27$

9 $y - (9 \div 3) = 19$

10 $b - 0.5 = 5.9$

11 $x - \dfrac{1}{3} = 3\dfrac{2}{3}$

12 $^{-}4 + c = 0$

Standards–Based Math • 5–6 © 2004 Creative Teaching Press

Inverse Operations: Multiplication

ALGEBRA

Use division to isolate the variable when multiplication is indicated.
Divide both sides by the same value.

$$30 = 4x - 2$$
$$32 = 4x$$
$$\frac{32}{4} = \frac{4x}{4}$$
$$8 = x$$

Solve.

1 $6x = 30$

2 $9x = 90$

3 $45 = 5a$

4 $65 = 3x + 5$

5 $27 = 6c + 3$

6 $7x = 42$

7 $40 = 5b + 5$

8 $28 = 4n$

9 $8x = 72$

10 $3x = 45$

11 $100a = 10,000$

12 $x^2 = 25$

Inverse Operations: Division

Algebra

Use multiplication to isolate the variable when division is indicated.
Multiply both sides by the same value.

$$32 = \frac{x}{2}$$

$$32 \times 2 = \frac{x}{2} \times 2$$

$$64 = x$$

Solve.

1 $\dfrac{a}{7} = 6$

2 $x \div 12 = 4$

3 $x \div 8 = 6$

4 $b \div 5 = 9$

5 $\dfrac{x}{4} = 5$

6 $n \div 10 = 12$

7 $\dfrac{x}{3} = 9$

8 $\dfrac{x}{5} + 6 = 21$

9 $3c \div 2 = 9$

10 $\dfrac{x}{8} + 9 = 20$

11 $2x \div 5 = 10$

12 $\dfrac{a}{5} - 8 = 3$

Variables 1

Algebra

Solve. Write the letter of each equation above its solution.

E $3(d + 4) = 27$

L $(6x)(5) = 60$

O $y + 3 = 15$

H $4 \times c + 15 = 39$

E $\dfrac{4x}{4} = 7$

F $6(3y) = 18$

R $10^x = 1{,}000$

Y $^-8(^-5x) = 800$

T $\dfrac{(3)(8)}{y} = 6$

N $\dfrac{1}{2}n = 50$

O $y^2 = 81$

Whose face launched a thousand ships?

__ __ __ __ __   __ __   __ __ __ __
6   7   2   5 100   12   1   4   3   9   20

Variables 2

Algebra

Solve. Write the letter of each equation above its solution.

A $3(7x) = 105$

A $\dfrac{(8)(4)}{y} = 16$

E $7 = 14 + x$

E $80 = x(4 + 6)$

G $3x = 45$

L $28 = 39 + x$

R $0 = 5x$

T $x + 40 = 102$

X $x - 6 = 45$

A $7 = 3 + x$

D $8 \times c - 8 = 48$

E $y^2 = 36$

E $\dfrac{6x}{3} = 20$

H $\dfrac{1}{2}n = 70$

N $90 = x + 18$

R $10^x = 0.0001$

T $y + 3 = 15$

For whom was Alexandria, Egypt named?

___	___	___	___	___	___	___	___	___
4	⁻11	6	51	5	72	7	⁻7	0

___	___	___		___	___	___	___	___
12	140	8		15	⁻4	10	2	62

Standards–Based Math • 5–6 © 2004 Creative Teaching Press

Variables 3

ALGEBRA

Solve. Write the letter or number of each equation above its solution.

U $35 - (13x) = 22$

S $(4x) + 15 = 43$

R $5 \cdot 10^x = 0.005$

N $\frac{1}{5}n + 6 = 9$

I $34 - x = 18$

I $\frac{3y}{-5} = 6$

1 $y^2 = 36$

O $72 = 8x$

S $y + 35 = 56$

S $76 - x = 28$

O $\frac{55 - x}{4} = 13$

M $6(d + 4) = 54$

I $(^-5x)^-6 = 300$

8 $26 \div x\ ^-6 = {}^-4$

O $\frac{x}{16} = {}^-4$

The three largest measured earthquakes in North America occurred in this state.

| ___ | ___ | ___ | ___ | ___ | ___ | ___ | ___ |
| 5 | 16 | 21 | 7 | 3 | 1 | ⁻3 | 10 |

| ___ | ___ | | ___ | ___ | ___ | ___ | ___ |
| ⁻10 | 15 | | 6 | 13 | ⁻64 | 9 | 48 |

Name _____ Date _____

Variables 4

Algebra

Solve. Write the letter of each equation above its solution.

A $7y + 3 = 24$

A $12(3y) = 288$

D $12y + 15 = 147$

D $^-4(3x) = 48$

E $^-5 = \dfrac{x}{5}$

H $\dfrac{3(6 + 11)}{y} = {^-}8\dfrac{1}{2}$

N $8.2 \times 10^x = 82{,}000$

O $(8x)(5) = 80$

O $\dfrac{6}{10}\, n = \dfrac{3}{4}$

R $7(y - 4) = 77$

S $\dfrac{6x}{2} = 21$

S $\dfrac{x}{48} = \dfrac{1}{4}$

S $\left(\dfrac{y}{6}\right)\left(\dfrac{2}{3}\right) = \dfrac{^-10}{18} = \dfrac{^-5}{9}$

U $y^3 = {^-}8$

The Romans invented concrete, which they used to build these.

$\overline{}$ $\overline{}$ $\overline{}$ $\overline{}$ $\overline{}$ $\overline{}$ $\overline{}$ $\overline{}$ $\overline{}$ $\overline{}$ $\overline{}$ $\overline{}$ $\overline{}$ $\overline{}$

15   2   3   11   7   8   4 $^-4$ 6 $\dfrac{5}{4}$ $^-2$ 12 $^-25$ $^-5$

Standards–Based Math • 5–6 © 2004 Creative Teaching Press

Evaluating Expressions

ALGEBRA

When the values of the variables are known, substitute the value for each variable and solve.

If x = 3 and y = 2:
4x + y =
(4)(3) + 2 =
12 + 2 = 14

Solve. a = 3, b = $\frac{1}{3}$, c = 5. Write your answer in simplest terms.

1 2c

2 4b

3 8a + c

4 9b + c

5 6ab

6 7c + 6a

Solve. a = 12, b = $\frac{3}{5}$, c = 9. Write your answer in simplest terms.

7 b$\frac{c}{a}$

8 ⁻6a − c

9 a − c

10 10(a + b) (Hint: Use distributive property!)

Solve.

11 x − $\frac{1}{3}$ = 3$\frac{2}{3}$

12 ⁻4 + c = 0

Inequalities

ᴀʟɢᴇʙʀᴀ

$7x > 49$

$\dfrac{7x}{7} > \dfrac{49}{7}$

$x > 7$

Use inverse operations to isolate the variable on one side of the expression. Whatever you do to one side of the expression must also be done to the other.

$x - 3 < 9$

$x - 3 + 3 < 9 + 3$

$x < 12$

Solve.

1 $3x - 7 < 2$

2 $-\dfrac{1}{3}x < 48$

3 $x + \dfrac{5}{6} < \dfrac{-1}{2}$

4 $x + 37 > 98$

5 $x - 53 < 141$

6 $\dfrac{x}{3} > \dfrac{1}{3}$

7 $\dfrac{11}{2}x < 3\dfrac{2}{3}$

8 $\dfrac{x}{-5} > 11$

9 $\dfrac{-7}{8}t > \dfrac{-7}{8}$

10 $\dfrac{x}{4} < -45$

11 $x + \dfrac{5}{8} > \dfrac{3}{4}$

12 $(18 - 12)x > 96$

Standards–Based Math • 5–6 © 2004 Creative Teaching Press

Functions 1

ALGEBRA

Complete.

$3x = y$

x	y
2	
4	
6	
8	
10	
12	

$4x - 8 = y$

x	y
3	
6	
9	
12	
15	
18	

$\dfrac{3x}{6} = y$

x	y
1	
3	
5	
7	
9	
11	

$5x + 3 = y$

x	y
2	
3	
4	
5	
6	
7	

$10^x = y$

x	y
⁻3	
⁻2	
⁻1	
0	
2	
4	

$\dfrac{x + 2}{x} = y$

x	y
8	
9	
10	
11	
12	
13	

Functions 2

Algebra

Complete.

$3(x + 4) = y$

x	y
2	
4	
6	
8	
10	
12	

$4(x - 8) = y$

x	y
3	
6	
9	
12	
15	
18	

$\dfrac{4x}{8} = y$

x	y
1	
3	
5	
7	
9	
11	

$6x + 7 = y$

x	y
2	
3	
4	
5	
6	
7	

$10^x = y$

x	y
-5	
-4	
-3	
1	
3	
5	

$\dfrac{x + 5}{x} = y$

x	y
8	
9	
10	
11	
12	
13	

Standards–Based Math • 5-6 © 2004 Creative Teaching Press

Functions 3

ᴀʟɢᴇʙʀᴀ

Complete.

$\dfrac{x}{6} = y$

x	y
2	
4	
6	
8	
10	
12	

$0.6x + 0.6 = y$

x	y
3	
6	
9	
12	
15	
18	

$\dfrac{^-5}{3}x + 4 = y$

x	y
1	
3	
5	
7	
9	
11	

$^-1.5x + 2.5 = y$

x	y
2	
3	
4	
5	
6	
7	

$4x + 8 = y$

x	y
$^-5$	
$^-4$	
$^-3$	
1	
3	
5	

$\dfrac{2}{3}x + \dfrac{1}{3} = y$

x	y
8	
9	
10	
11	
12	
13	

Writing and Solving Equations

ALGEBRA

Write an equation that illustrates the problem using *n* as the variable. Solve. There may be more than one right way to write the equation.

1 Haley placed crayons in six equal piles. Then she called the preschoolers to the table for art time. "Will that be enough?" asked one of the other preschool teachers. "Oh, sure!" Haley replied. "There are 60 altogether." How many crayons were in each pile?

2 Nancy placed 144 candies on the table. "Hmm," she wondered aloud. "There are twelve children. How many do I place in each party bag so that they all have the same amount?" How many candies should Nancy put in each bag?

3 It is pet adoption day at Pets Inc. The local animal rescue groups arrive and begin to set up. Austin counts each kind of animal. There are 3 rabbits, some kittens, 12 dogs, and a parrot awaiting adoption. "How many animals do you have today?" asks the store manager. "Twenty-two altogether," Austin replies. How many of the animals are kittens?

4 For a food co-op, a small group of people order a large amount of food and share it. Tonight the Centerville Food Co-op meets to divide up 5,000 lbs. of bread flour. Each family leaves with 250 lbs. of bread flour. How many families participated in the co-op?

5 The Jacobs family has been on the road with their RV for four days. They've driven the same distance on each day for a total of 1,660 miles. How many miles have they driven each day?

6 Sam rides 15 miles to school, 8 miles to his grandma's house after school, and then he rides home for dinner. Altogether, he is riding more than 40 miles every day. What do you know about the distance Sam rides from his grandmother's house to his home?

Name _____ Date _____

Venn Diagrams

ALGEBRA

John Venn invented the Venn diagram as a method of visualizing logical relationships. A Venn diagram has two or more circles. Each circle represents a specific group. If the sets share common values, then the circles overlap and those values are written in the shared area.

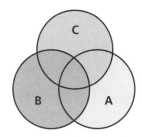

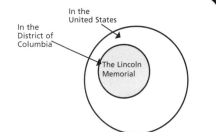

In the District of Columbia

In the United States

The Lincoln Memorial

Read the problem. Complete the Venn diagram to solve it. The first problem has been started for you.

1 Mrs. Hooper's three cats had kittens and she needs to find them all homes. She makes a list of the characteristics of the 24 kittens.
- 10 kittens are black
- 8 kittens have white ears
- 16 kittens are long-haired
- One kitten is black and has white ears and long hair
- Three kittens have white ears and long hair but are not black
- Two kittens are black with white ears, but they do not have long hair

Her friend offers to adopt the kittens that are black with long hair, but do not have white ears. How many kittens will she adopt?

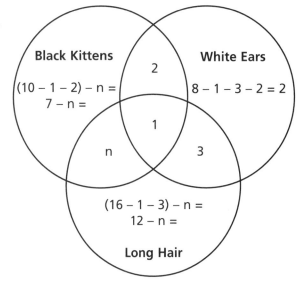

Black Kittens

$(10 - 1 - 2) - n =$
$7 - n =$

2

White Ears

$8 - 1 - 3 - 2 = 2$

1

n

3

$(16 - 1 - 3) - n =$
$12 - n =$

Long Hair

$2 + 1 + 3 + n +$ _____ $+$ _____ $+ 2 = 24$
$n =$ _____

2 There are 100 students in the 6th grade.
- 60 take Spanish
- 25 take Latin
- 40 take French
- There is one student enrolled in all three courses.
- 2 students are enrolled in both French and Spanish.
- 15 Latin students are enrolled in Spanish.

How many Latin students are enrolled in French?

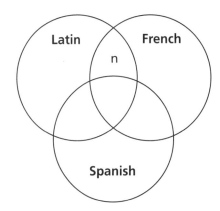

Latin

French

n

Spanish

Standards-Based Math • 5–6 © 2004 Creative Teaching Press

Name _____ Date _____

Angles

Geometry

Match each illustration to its definition.

_____ **1**

_____ **2**

_____ **3**

_____ **4**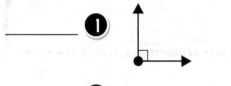

_____ **5**

_____ **6**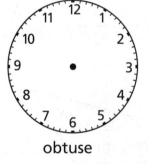

a. Acute Angle: measures less than 90°

b. Right Angle: measures exactly 90°

c. Supplementary Angle: two angles that create a 180° angle

d. Obtuse Angle: measures more than 90°

e. Complementary Angle: two angles that create a 90° angle

f. Straight Angle: measures exactly 180°

Draw hands on the clock that illustrate the kind of angle indicated.

7

straight

8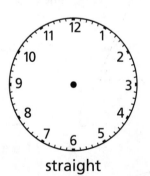

obtuse

9

acute

Standards-Based Math • 5–6 © 2004 Creative Teaching Press

Angle Measurement

GEOMETRY

The sum of the interior angles of a triangle is 180°.
∠A + ∠B + ∠C = 180°

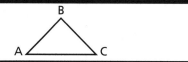

Find the measure of angle A.

1

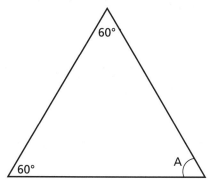

2

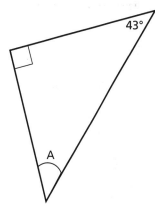

3

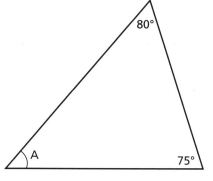

4

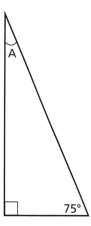

5

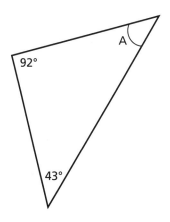

6

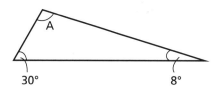

Name _____ Date _____

Geometry Terms

GEOMETRY

Write the word that completes each clue. Use these words to complete the crossword puzzle.

Across

2. A __ is a flat surface that extends endlessly in all directions.
3. Two rays joined at a common endpoint form an ___.
4. A __ is a flat surface of a solid.
5. The __ is a point equal distance from the endpoints of a line.
6. A __ is a geometric object that has no dimensions, only a location.
7. Part of a line with only one endpoint is called a ___.
8. Two faces of a solid meet at the __.
9. A __ is a three-dimensional geometric figure.
10. Lines that never intersect and are in the same plane are called __.

Down

1. ___ lines meet at one point.
6. Lines that intersect at a right angle are __ lines.

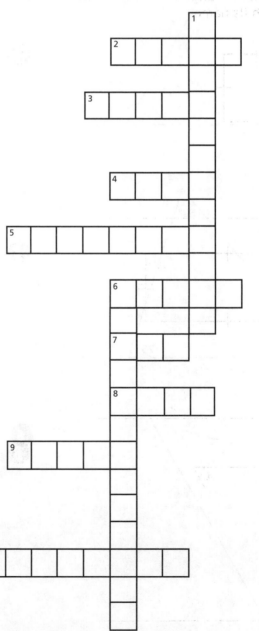

Standards–Based Math • 5–6 © 2004 Creative Teaching Press

Quadrilaterals

GEOMETRY

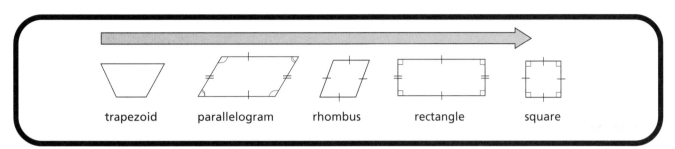

trapezoid   parallelogram   rhombus   rectangle   square

The sum of the interior angles of a quadrilateral is 360°. Find the value of angle A and label each quadrilateral with its name.

1

∠A = _____ _____

2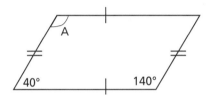

∠A = _____ _____

3

∠A = _____ _____

4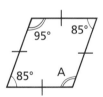

∠A = _____ _____

5

∠A = _____ _____

6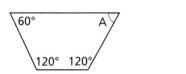

∠A = _____ _____

Name _____   Date _____

Polygons
Geometry

A polygon's prefix names its number of sides. Most polygon prefixes are Greek in origin.

Number	=	Prefix	Polygon
treis, tria	3	tri-	triangle
quattor*	4	quadri-, quart-	quadrilateral
pente	5	penta-	pentagon
hexa	6	hex-	hexagon
hepta	7	hept-	heptagon
okto	8	oct-	octagon
ennea	9	ennea	nonagon
deka	10	dec-	decagon
dodeka	12	dodec-	dodecagon

*Latin in origin, as in *lateral,* which means "sides."

Label each polygon.

1

_____

2

_____

3

_____

4

_____

5

_____

6

_____

7

_____

8

_____

9

_____

10

_____

11

_____

12

_____

Standards-Based Math • 5-6 © 2004 Creative Teaching Press

Similar and Congruent

GEOMETRY

Similar shapes are the same shape but not the same size.   **Congruent** shapes are the same shape and size.

Label each set of figures **similar** or **congruent**.

1

_____

2

_____

3

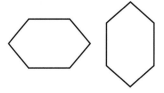

_____

4

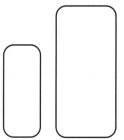

_____

5

_____

6

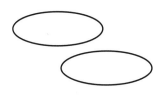

_____

7

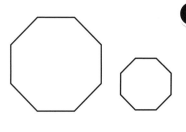

_____

8

_____

9

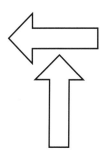

_____

Name _____ Date _____

Similar Triangles

Geometry

Similar triangles have congruent angles. The perimeters and each corresponding side share the same ratio. This is called **the ratio of scale.**

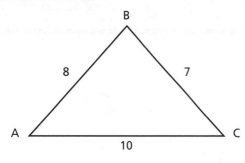

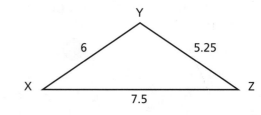

$$\frac{\overline{AB}}{\overline{XY}} = \frac{8}{6} = 1\frac{1}{3} = 1.\overline{33}$$ $$\frac{\overline{BC}}{\overline{YZ}} = \frac{7}{5.25} = 1.\overline{33}$$ $$\frac{\overline{AC}}{\overline{XZ}} = \frac{10}{7.5} = 1.\overline{33}$$

$\triangle ABC \sim \triangle XYZ \sim \triangle HIJ$

Use a known ratio to find the length of unknown sides.

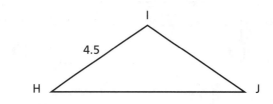

$$\frac{\overline{YZ}}{\overline{IJ}} = \frac{5.25}{1.33} = 3.9375 \approx 3.9$$ $$\frac{\overline{XZ}}{\overline{HJ}} = \frac{7.5}{1.33} = 5.625 \approx 5.6$$

$$\overline{IJ} = 3.9 \qquad \overline{HJ} = 5.6$$

Evaluate the similar triangles. Find the ratio of scale. Use this information to fill in all the missing side measurements. Use what you know about triangles to fill in the missing angle measurements.

 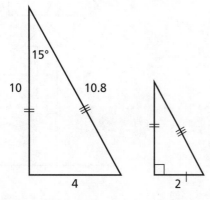

 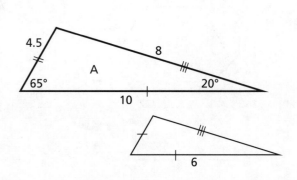

Standards-Based Math • 5–6 © 2004 Creative Teaching Press

Similar Figures

GEOMETRY

Similar figures also have equal angles and sides that share the same scale ratio.

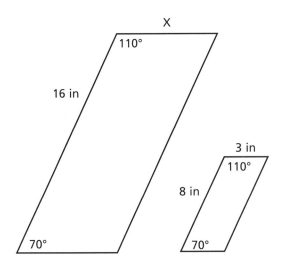

$$\frac{16}{x} = \frac{8}{3}$$

$$8x = 3 \times 16$$

$$\frac{8x}{8} = \frac{48}{8}$$

$$x = 6 \text{ in}$$

Calculate the ratio and complete the missing side measurements.

1

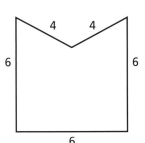

2

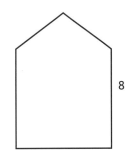

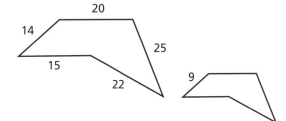

3

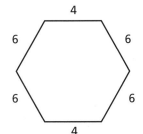

4

Name _____ Date _____

Prisms and Pyramids

Geometry

Prisms are solids with two parallel, congruent polygon bases and rectangular faces that connect them.

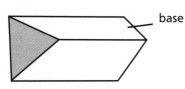

base

Pyramids are solids with a polygon base and triangular faces that come to a point.

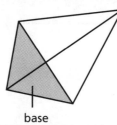

base

Name the shape. Label the type of polygon of the base(s).

  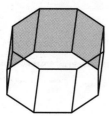

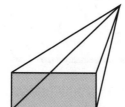

_____

_____

_____

_____

_____

_____

  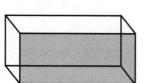

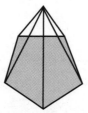

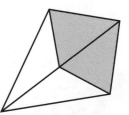

_____

_____

_____

_____

_____

_____

Standards-Based Math • 5-6 © 2004 Creative Teaching Press

Cylinders, Cones, and Spheres

Geometry

Cylinders are solids with two circular bases and a curved surface.

Cones are solids with a circular base, a curved surface, and one vertex.

Spheres have a curved surface where every point is an equal distance from the center.

Look at the everyday object. Name the polyhedron it resembles.

1

_____

2

_____

3

_____

4

_____

5

_____

6

_____

7

_____

8

_____

9

_____

Pythagorean Proof

GEOMETRY

Pythagoras (~569 B.C.–475 B.C.) was a Greek mathematician. The rule that helps you find the hypotenuse of a right triangle is named for him.

$$a^2 + b^2 = c^2$$

Prove the theory for yourself. Cut out square a and glue it on square c. Cut out square b and glue it on square c. Hint: Cut square b into smaller pieces along the dotted lines.

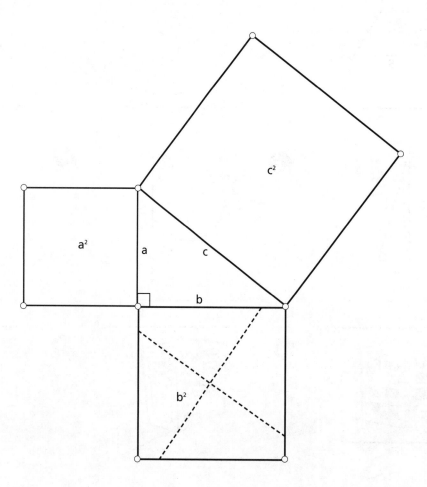

Pythagorean Theorem

Geometry

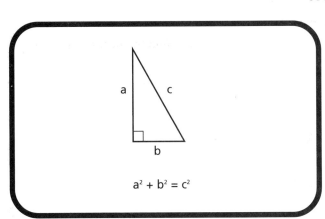

$$a^2 + b^2 = c^2$$

Use a calculator with a $\sqrt{}$ button to help you find the hypotenuse of each triangle.

1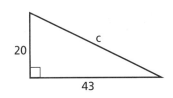

_____

2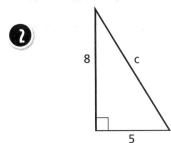

_____

3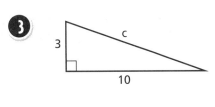

_____

4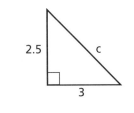

_____

5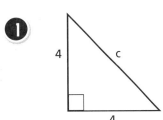

_____

6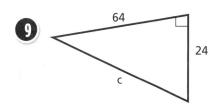

_____

7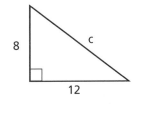

_____

8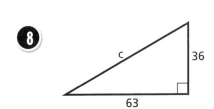

_____

9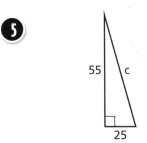

_____

10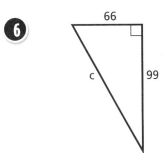

_____

Parts of a Circle

GEOMETRY

Match each word to its definition.

_____ **1** radius

_____ **2** chord

_____ **3** tangent

_____ **4** diameter

_____ **5** arc

_____ **6** semi-circle

a. A line segment with both endpoints on the circle and that passes through the center.

b. An arc that is one-half the circumference of a circle.

c. The distance from the center to a point on the circle.

d. A line perpendicular to the radius that touches only one point on the circle.

e. A line segment that connects two points on a curve.

f. A portion of the circumference of a circle.

Use colored pencils to mark the circle as directed.

Use a red pencil to draw radius $\overline{AB}$.

Use a green pencil to draw chord $\overline{CD}$.

Use a purple pencil to show the diameter $\overline{EAF}$.

Use a yellow pencil to trace the arc $\overparen{CD}$.

Use an orange pencil to trace the semi-circle $\overparen{EBF}$.

Use a blue pencil to show tangent $\overleftrightarrow{GH}$.

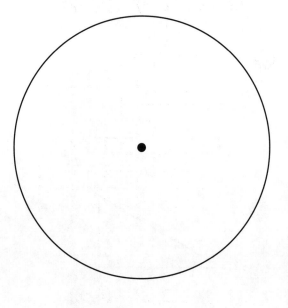

Standards–Based Math • 5–6 © 2004 Creative Teaching Press

The X,Y Axis

GEOMETRY

The **Cartesian plane** is made up of a horizontal line called the **X-axis** and a vertical line called the **Y-axis.** They are perpendicular to each other and meet at point zero.

Any point on the plane can be described with a pair of numbers called an **ordered pair.** The first number always refers to where the point lines up with the X-axis and the second number always refers to where the point lines up with the Y-axis.

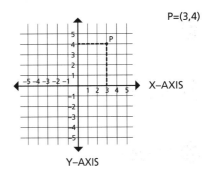

P=(3,4)

Circle the correct set of coordinates for Point P.

1

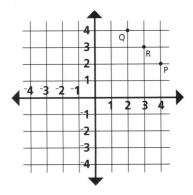

(3, 2)   (2, 3)   (3, –2)

2

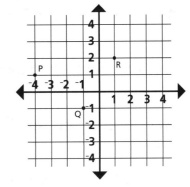

(2, –2)   (2, 3)   (3, –2)

3

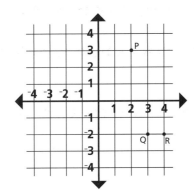

(1, 2)   (3, 3)   (4, 2)

4

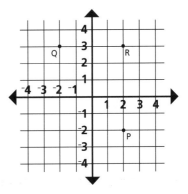

(1, 2)   (–4, 1)   (–1, –1)

Coordinate Pairs

GEOMETRY

Write the ordered pair for each point.

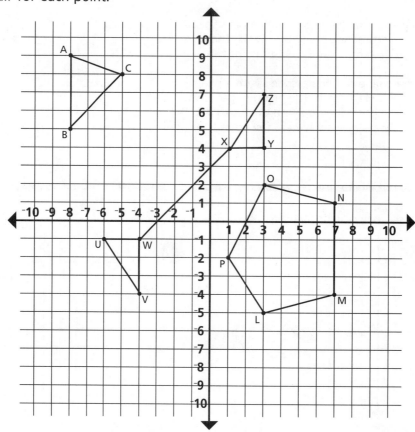

1 A =

2 B =

3 C =

4 L =

5 M =

6 N =

7 O =

8 P =

9 U =

10 V =

11 W =

12 X =

13 Y=

14 Z =

Standards–Based Math • 5–6 © 2004 Creative Teaching Press

Name _____ Date _____

Using a Cartesian Plane

Geometry

Write the ordered pair at which you stop.

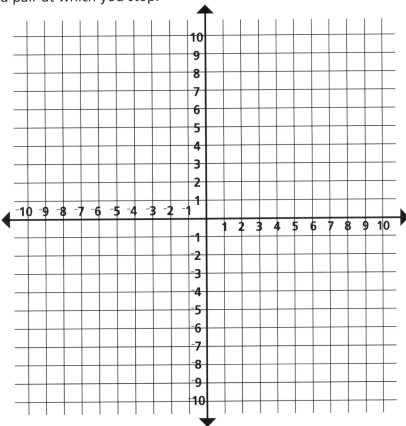

1 Start at point (2, 3). Move 2 points to the right, 2 to the left, and 3 to the left. Final point: _____

2 Start at point (5, 4). Move 4 points to the right, 2 to the left, and 8 to the left. Final point: _____

3 Start at point (6, ⁻2). Move 8 points to the left, and 5 up. Final point: _____

4 Start at point (2, 8). Move 2 points up, 1 down, and 3 to the right. Final point: _____

5 Start at point (⁻3, ⁻5). Move 0 points to the left, 4 down, and 5 to the right. Final point: _____

6 Start at point (6, 3). Move 0 points down, 2 up, and 5 down. Final point: _____

7 Start at point (⁻5, 0). Move 2 points to the right, 9 up, and 5 down. Final point: _____

8 Start at point (⁻3, 4). Move 8 points to the left, 5 to the right, and 7 points up. Final point: _____

Standards-Based Math • 5-6 © 2004 Creative Teaching Press

Plotting Points

GEOMETRY

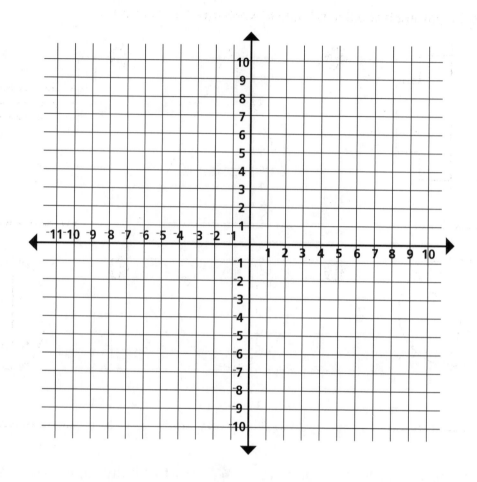

Plot each point on the plane. Connect each point with a line segment in the order given.

1.	2.	3.
(⁻9, ⁻2)	(4, 2)	(⁻4, 6)
(⁻11, ⁻4)	(4, ⁻4)	(⁻4, 3)
(⁻11, ⁻7)	(6, ⁻4)	(⁻7, 3)
(⁻9, ⁻9)	(6, 2)	(⁻6, 4)
(⁻6, ⁻9)	(8, 2)	(⁻10, 8)
(⁻4, ⁻7)	(8, 4)	(⁻9, 8)
(⁻4, ⁻4)	(2, 4)	(⁻9, 9)
(⁻6, ⁻2)	(2, 2)	(⁻5, 5)
(⁻9, ⁻2)	(4, 2)	(⁻4, 6)

Standards–Based Math • 5–6 © 2004 Creative Teaching Press

Reflection Symmetry 1

GEOMETRY

Write **yes** or **no** to indicate if the line is a line of symmetry for that figure.

1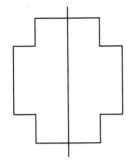

_____

2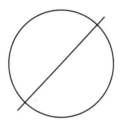

_____

3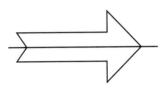

_____

4

_____

5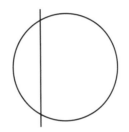

_____

6

_____

Complete the image to create a symmetrical figure.

7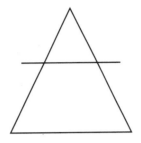

8

9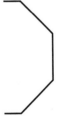

Standards-Based Math • 5–6 © 2004 Creative Teaching Press

Reflection Symmetry 2

GEOMETRY

Draw a line of symmetry for each figure.

1

2

3

4

5

6

7

8

9

10

11

12

Standards-Based Math • 5–6 © 2004 Creative Teaching Press

Rotational Symmetry

Geometry

Rotational symmetry is when a figure can be rotated around its central point and look the same in a position other than the one in which it started.

no rotational symmetry

rotational symmetry

Write **yes** if the figure has rotational symmetry. Write **no** if it does not.

1

_____

2

_____

3

_____

4

_____

5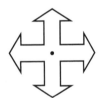

_____

6

_____

7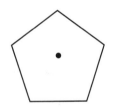

_____

8

_____

9

_____

10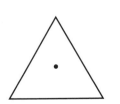

_____

11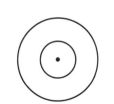

_____

12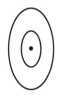

_____

Reflections Across the X-Axis

GEOMETRY

When a figure is reflected across the X-axis, the x coordinates do not change. The y coordinates switch from + to – or – to +.

(2, 3)   (2, ⁻3)
(3, 5)   (3, ⁻5)
(4, 4)   (4, ⁻4)

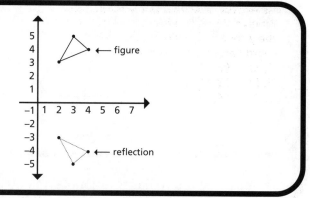

Record the reflected coordinates on the table. Then draw an XY axis on graph paper and check your work by plotting both sets of points.

①

Figure	Reflection
(2, ⁻8)	
(2, ⁻2)	
(4, ⁻2)	
(4, ⁻6)	
(7, ⁻6)	
(7, ⁻8)	

②

Figure	Reflection
(1, 3)	
(1, 8)	
(5, 8)	
(5, 6)	
(4, 6)	
(4, 5)	
(5, 5)	
(5, 3)	

③

Figure	Reflection
(⁻3, 3)	
(⁻5, 6)	
(⁻3, 9)	
(⁻5, 9)	
(⁻6, 7)	
(⁻7, 9)	
(⁻9, 9)	
(⁻7, 6)	
(⁻9, 3)	
(⁻7, 3)	
(⁻6, 5)	
(⁻5, 3)	

④

Figure	Reflection
(⁻6, ⁻2)	
(⁻4, ⁻4)	
(⁻6, ⁻5)	
(⁻4, ⁻8)	
(⁻8, ⁻8)	
(⁻10, ⁻5)	
(⁻8, ⁻4)	
(⁻10, ⁻2)	

Standards–Based Math • 5–6 © 2004 Creative Teaching Press

Name _____ Date _____

Reflections Across the Y-Axis

GEOMETRY

When a figure is reflected across the Y-axis, the y coordinates do not change. The x coordinates switch from + to – or – to +.

(2, 3) (⁻2, 3)
(3, 5) (⁻3, 5)
(4, 4) (⁻4, 4)

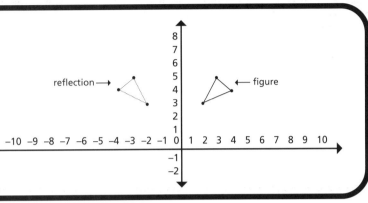

reflection →   ← figure

Record the reflected coordinates on the table. Then draw an XY axis on graph paper and check your work by plotting both sets of points.

①

Figure	Reflection
(⁻2, 6)	
(⁻2, 8)	
(⁻5, 8)	
(⁻5, 7)	
(⁻7, 7)	
(⁻7, 4)	
(⁻5, 4)	
(⁻5, 6)	

②

Figure	Reflection
(⁻3, 3)	
(⁻3, 6)	
(⁻4, 5)	
(⁻4, 8)	
(⁻7, 7)	
(⁻7, 5)	
(⁻6, 6)	
(⁻6, 3)	
(⁻7, 2)	
(⁻4, 2)	

③

Figure	Reflection
(2, 4)	
(2, 6)	
(4, 9)	
(5, 7)	
(8, 9)	
(8, 6)	
(5, 6)	
(6, 4)	

④

Figure	Reflection
(4, 5)	
(6, 7)	
(9, 7)	
(11, 5)	
(8, 5)	
(10, ⁻2)	
(5, ⁻2)	
(7, 5)	

Name _____ Date _____

Translations Along the X-Axis

GEOMETRY

Adding or subtracting a constant value to the X-axis of each point of a figure will translate it along the X-axis.

x, y	(x – 6), y
(3, 4)	(‾3, 4)
(3, 7)	(‾3, 7)
(5, 6)	(‾1, 6)
(7, 7)	(1, 7)
(7, 4)	(1, 4)
(5, 5)	(‾1, 5)

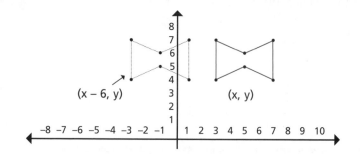

Record the translated coordinates on the table. Then, draw an XY axis on graph paper and check your work by plotting both sets of points.

①

x + 7	
Figure	**Translation**
(‾9, 4)	
(‾6, 8)	
(‾5, 4)	

②

x + 5	
Figure	**Translation**
(5, 5)	
(6, 10)	
(8, 8)	
(8, 5)	

③

x – 4	
Figure	**Translation**
(3, ‾9)	
(7, ‾3)	
(8, ‾6)	
(6, ‾9)	

④

x + 8	
Figure	**Translation**
(‾6, 2)	
(‾7, 4)	
(‾6, 6)	
(‾7, 8)	
(‾6, 10)	
(‾4, 10)	
(‾5, 8)	
(‾4, 6)	
(‾5, 4)	
(‾4, 2)	

Standards–Based Math • 5–6 © 2004 Creative Teaching Press

Translations Along the Y-Axis

Geometry

Adding or subtracting a constant value to the Y-axis of each point of a figure will translate it along the Y-axis.

x, y	x, (y – 7)
(7, 3)	(7, ⁻4)
(5, 5)	(5, ⁻2)
(5, 9)	(5, 2)
(8, 7)	(8, 0)
(8, 5)	(8, ⁻2)

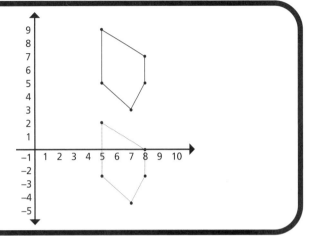

Record the translated coordinates on the table. Then draw an XY axis on graph paper and check your work by plotting both sets of points.

1

y – 5	
Figure	**Translation**
(2, 5)	
(2, 7)	
(8, 5)	
(7, 4)	
(4, 4)	

2

y + 6	
Figure	**Translation**
(⁻10, ⁻3)	
(⁻9, ⁻7)	
(⁻7, ⁻6)	
(⁻5, ⁻4)	
(⁻8, ⁻4)	

3

y – 10	
Figure	**Translation**
(3, 5)	
(3, 7)	
(5, 8)	
(7, 7)	
(9, 8)	
(9, 5)	

4

y + 8	
Figure	**Translation**
(⁻10, ⁻2)	
(⁻7, ⁻2)	
(⁻5, ⁻4)	
(⁻3, ⁻2)	
(⁻3, ⁻7)	
(⁻5, ⁻5)	
(⁻7, ⁻7)	
(⁻10, ⁻7)	

Name _____ Date _____

Oblique Translations

Geometry

Adding or subtracting a constant value to the X-axis
and Y-axis of each point of a figure will translate it
to another location on the grid.

x, y	(x – 4), (y – 7)
(7, 3)	(3, –4)
(5, 5)	(1, –2)
(5, 9)	(1, 2)
(8, 7)	(4, 0)
(8, 5)	(4, –2)

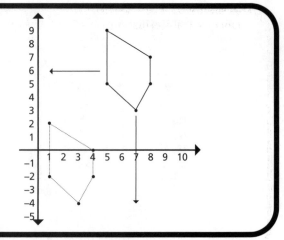

Record the translated coordinates on the table. Then draw an XY axis on graph paper and check your
work by plotting both sets of points.

1

x + 5, y – 5	
Figure	**Translation**
(2, 4)	
(2, 5)	
(6, 5)	
(6, 4)	

2

x – 7, y + 6	
Figure	**Translation**
(1, 3)	
(1, 7)	
(3, 5)	
(3, 2)	

3

x – 3, y – 10	
Figure	**Translation**
(12, 5)	
(9, 3)	
(12, 2)	
(15, 3)	

4

x + 8, y + 8	
Figure	**Translation**
(⁻4, ⁻2)	
(⁻2, ⁻4)	
(⁻3, ⁻5)	
(⁻2, ⁻7)	
(⁻4, ⁻7)	
(⁻5, ⁻5)	
(⁻4, ⁻4)	
(⁻6, ⁻2)	

Standards-Based Math • 5–6 © 2004 Creative Teaching Press

Transformations

Geometry

Oblique Transformation

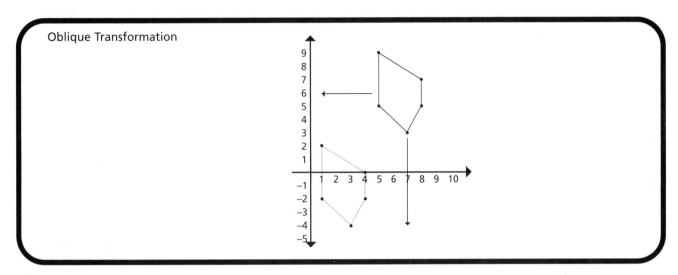

Look at the diagram. Name the kind of transformation performed.

1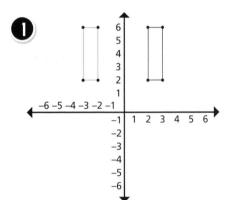

_____

2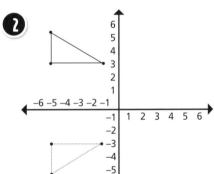

_____

3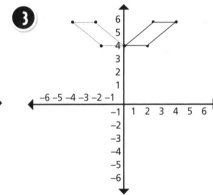

_____

4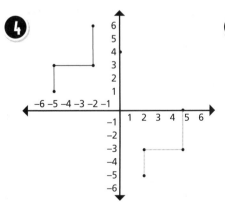

_____

5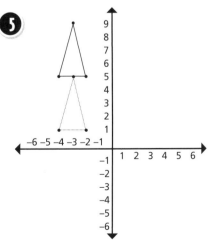

_____

6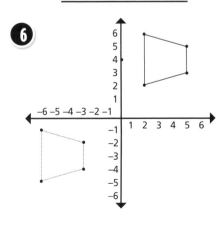

_____

Name _____ Date _____

Perimeter

MEASUREMENT

Perimeter is the measurement of the distance around an object.

Find the perimeter.

1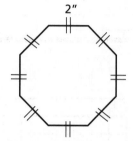

_____

2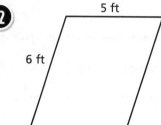

_____

3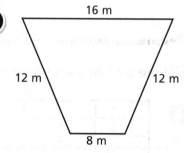

_____

4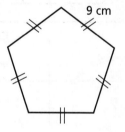

_____

5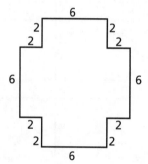

_____

6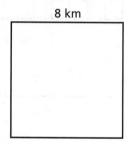

_____

7

_____

8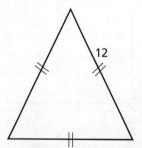

_____

9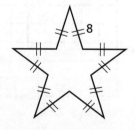

_____

Standards-Based Math • 5–6 © 2004 Creative Teaching Press

Area

MEASUREMENT

Area is the measurement of the total surface of the object. Count the squares contained in the figure.

area = 18 squares units

Find the area of each figure by counting square units.

1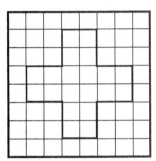

_____

2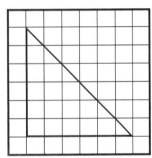

_____

3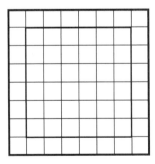

_____

4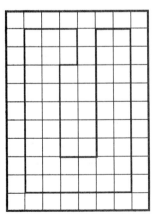

_____

5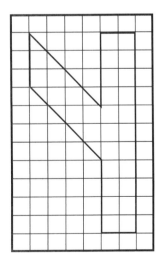

_____

6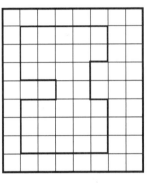

_____

Area of a Triangle

Measurement

Area = $\frac{1}{2}$ bh

8

6

Area = $\frac{1}{2}$ (6)(8) = 24

Find the area of each triangle.

1

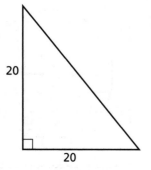

20

20

_____

2

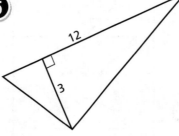

16

15

_____

3

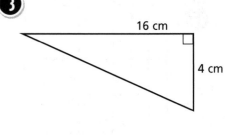

16 cm

4 cm

_____

4

20

60

_____

5

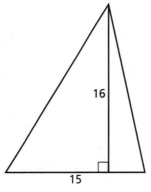

12

3

_____

6

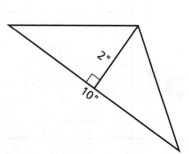

2"

10"

_____

Name _____ Date _____

Circumference

The **circumference** of a circle is the distance around the outside of the circle. Circumference = the perimeter of a circle.

C = πd   d = diameter
π = 3.14   r = radius
C = πd   d = 2r
C = π(2)(9) = π18 = (3.14)(18) = 56.52

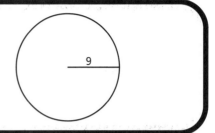

Find the circumference of each circle.

1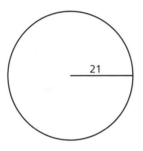
21

_____

2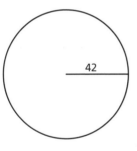
15

_____

3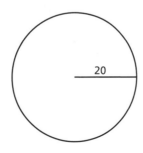
3

_____

4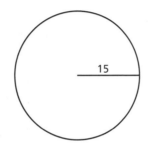
42

_____

5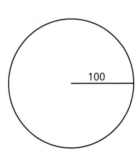
81

_____

6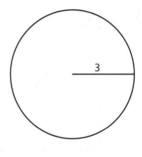
12

_____

7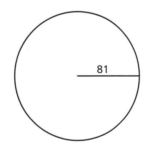
100

_____

8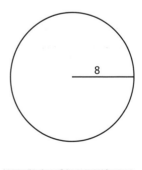
20

_____

9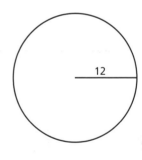
8

_____

Name _____ Date _____

Area of Circles

MEASUREMENT

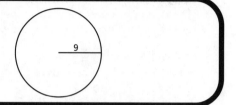

$A = \pi r^2$ $r = $ radius
$\pi = 3.14$
$A = \pi 9^2 = \pi 81 = 254.34$

Find the area of each circle.

1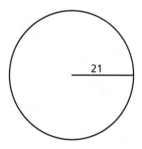
21

_____

2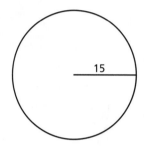
15

_____

3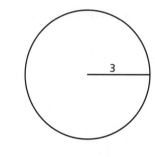
3

_____

4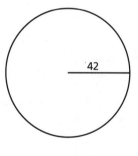
42

_____

5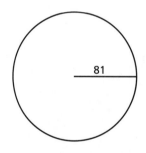
81

_____

6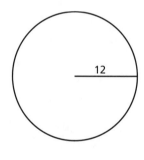
12

_____

7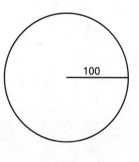
100

_____

8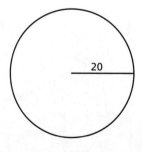
20

_____

9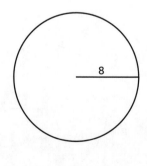
8

_____

Standards–Based Math • 5–6 © 2004 Creative Teaching Press

Name _____ Date _____

Circumference and Area of Circles

MEASUREMENT

$$A = \pi r^2 \qquad C = \pi d$$

Find the area and circumference of each circle.

1 A _____

C _____

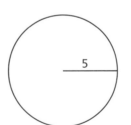

2 A _____

C _____

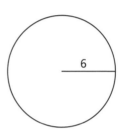

3 A _____

C _____

4 A _____

C _____

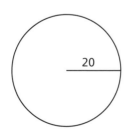

5 A _____

C _____

6 A _____

C _____

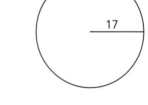

7 A _____

C _____

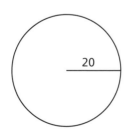

8 A _____

C _____

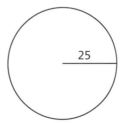

9 A _____

C _____

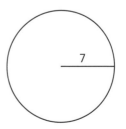

Standards-Based Math • 5–6 © 2004 Creative Teaching Press

Name _____ Date _____

Perimeter and Area of Curved Shapes

Measurement

$\square$ Area = bh   Perimeter = sum of the sides

$\triangle$ Area = $\frac{1}{2}$ bh   Circumference = πd

$\bigcirc$ Area = πr^2

Find the area and the perimeter of each figure.

1 A _____

   P _____

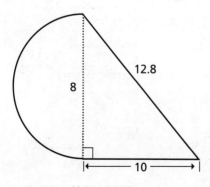

2 A _____

   P _____

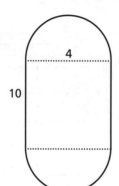

3 A _____

   P _____

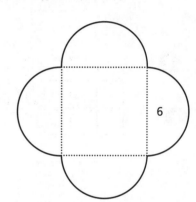

4 A _____

   P _____

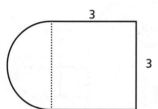

5 A _____

   P _____

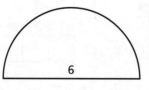

6 A _____

   P _____

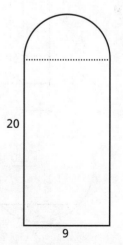

Standards-Based Math • 5–6 © 2004 Creative Teaching Press

Volume

MEASUREMENT

Find the volume of each figure by counting cubes.

1

2

3

_____   _____   _____

4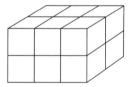

5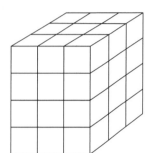

6

_____   _____   _____

7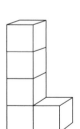

8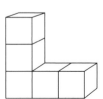

9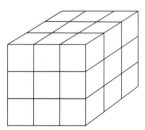

_____   _____   _____

Volume of Rectangular Prisms

Mᴇᴀsᴜʀᴇᴍᴇɴᴛ

Volume = length × width × height

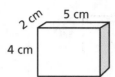

Volume = 5 cm × 2 cm × 4 cm = 40 cm³

Find the volume of each figure.

1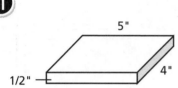

5"
1/2"
4"

_____

2

3 cm   5 cm

12 cm

_____

3

2 ft   18 ft

12 ft

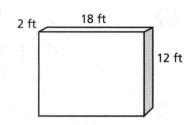

_____

4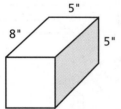

5"
8"
5"

_____

5

4 m
16 m
10 m

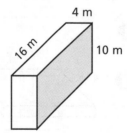

_____

6

3 cm   7 cm

9 cm

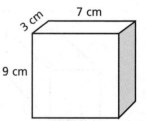

_____

Standards-Based Math • 5–6 © 2004 Creative Teaching Press

Surface Area

Measurement

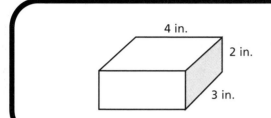

Surface area = the sum of the area of all surfaces
(4)(3) + (4)(3) + (2)(3) + (2)(3) +(2)(4) + (2)(4) =
24 + 12 + 16 = 52

Find the surface area.

1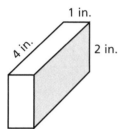

_____

2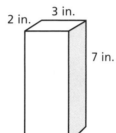

_____

3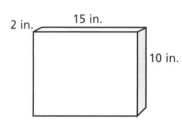

_____

4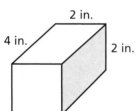

_____

5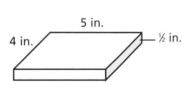

_____

6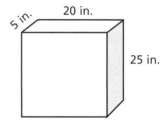

_____

Measurement Equivalents

MEASUREMENT

1 m = 39.37 in.	1 kg = 2.2 lb.	1 qt = 0.95 L
2.54 cm = 1 in.	1 oz = 28.35 g	1 gal = 3.79 L
1 km = 0.62 mi		

Use the chart to solve the word problems.

1 A photograph is 3 in. by 5 in. About how many centimeters wide and long is the photograph?

2 Mom buys 3 quarts of milk on her way home from work. Dad buys 3 L of milk on his way home from work. Who bought more milk?

3 A coleslaw recipe calls for 600 g of cabbage. About how many ounces of cabbage is needed?

4 Max's mom sends him to the store to buy a 10 lb. box of laundry detergent. To his surprise all the boxes list their weight in kg. How much must the box weigh to meet his mother's request?

5 Henry's dog is getting vaccinated. The vet has to change the dog's weight from pounds to kilograms to get the correct dose. If the dog is 15 lb., how much does it weigh in kg?

6 Karen wants to order a special toy for her son from a company in Canada. Karen knows most companies charge more to ship packages over 25 lb. The toy is listed as 16 kg. Will she most likely have an extra charge for shipping? Why or why not?

7 Ben and Milo are twins on the swim team. Ben swims a 1500 meter race. Milo swims the mile. Whose race is longer? By how much?

Name _____ Date _____

Length and Weight

MEASUREMENT

Length	Weight
12 in. = 1 ft.	16 oz. = 1 lb.
3 ft. = 1 yd.	2000 lb. = 1 T
5,280 ft. = 1 mi	

Use the chart to find the equivalencies.

1 5 yd. = ___ in.

2 6 ft. = ___ yd.

3 120 oz. = ___ lb.

4 4000 lb. = ___ T

5 4 ft. = ___ in.

6 8 oz. = ___ lb.

7 18 ft. = ___ yd.

8 4 oz. = ___ lb.

9 3 yd. = ___ ft.

10 12 oz. = ___ lb.

11 36 in. = ___ ft.

12 10 oz. = ___ lb.

13 1.5 T = ___ lb.

14 4 ft. = ___ yd.

15 3.5 lb. = ___ oz.

Standards-Based Math • 5–6 © 2004 Creative Teaching Press

Capacity

MEASUREMENT

> Knowing how to convert standard units of capacity is helpful for doubling or halving recipes or other directions (such as how much water is needed for a fish tank) when very large or very small quantities are involved.
>
> | | |
> |---|---|
> | 3 tsp = 1 Tbsp | 2 c = 1 pt. |
> | 16 tbsp = 1 c | 2 pt = 1 qt. |
> | 8 oz. = 1 c | 4 qt = 1 gal. |

Find the equivalent measurement.

1 8 qt. = ___ pt.

2 8 pt. = ___ qt.

3 5 gal. = ___ qt.

4 2 tbsp = ___ Tbsp

5 $\frac{1}{2}$ c = ___ Tbsp

6 $\frac{3}{4}$ c = ___ oz.

7 $\frac{1}{2}$ gal. = ___ qt.

8 8 c = ___ qt.

9 12 c = ___ gal.

10 4 Tbsp = ___ c

11 12 tsp= ___ c

12 12 oz. = ___ c

13 1 qt. = ___ oz.

14 4 oz. = ___ c

15 8 oz. = ___ pt.

Standards–Based Math • 5–6 © 2004 Creative Teaching Press

Average

DATA ANALYSIS AND PROBABILITY

3 cm ————————
6 cm ————————————
12 cm ——————————————————————
To calculate the average length of the strings, add all three lengths and divide the total by 3.

3 + 6 + 12 = 21
21 ÷ 3 = 7
The average length of the strings is 7 cm.

Find the average for each set of numbers.

1. 8 cm, 5 cm, 4 cm, 5 cm, 3 cm

2. 24, 15, 30

3. 10°C, 20°C, 30°C, 50°C

4. 4, 24, 16, 32

5. 54, 72, 27

6. 28¢, 56¢, 45¢, 67¢, 32¢, 45¢

7. 2, 5, 7, 4, 5, 3, 7, 8, 2, 4, 6, 7

8. 3.6 in., 6.3 in., 6.0 in.

9. 700, 1001, 898, 954, 1012, 987, 876, 568

10. 4.5, 6.7, 5.5, 8.2, 4.1, 6.5, 3.5, 5.6, 7.7, 9.0

Name _____   Date _____

Mean, Median, Mode, and Range

DATA ANALYSIS AND PROBABILITY

> **Mean:** the average
> **Median:** the middle number of the data when the data is arranged in order
> **Mode:** the data value that appears most frequently
> **Range:** the difference between the greatest and least value of the data set

Find the mean, median, mode, and range for each set of data.

1 6, 9, 8, 9, 12, 15, 12, 6, 9, 7, 15, 12

   mean :_____   median :_____   mode :_____   range :_____

2 0.43, 0.76, 0.46, 0.55, 0.25, 0.32, 0.37, 0.50

   mean :_____   median :_____   mode :_____   range :_____

3 89, 93, 51, 64, 91, 103, 46, 64, 82, 123, 112, 99

   mean :_____   median :_____   mode :_____   range :_____

4 16, 23, 18, 30, 19, 37, 23, 30, 33, 16, 30

   mean :_____   median :_____   mode :_____   range :_____

5 17,423; 13,678; 19,555; 18,000; 16,894

   mean :_____   median :_____   mode :_____   range :_____

6 $24.00, $160.00, $78.00, $44.00, $48.00, $36.00, $44.00

   mean :_____   median :_____   mode :_____   range :_____

Standards–Based Math • 5–6 © 2004 Creative Teaching Press

Analyzing Data

DATA ANALYSIS AND PROBABILITY

It's time for the Butter and Eggs Day Parade! Helen takes information from each group in the parade. She makes the following chart:

Group	No. of People Riding on Float	No. of Animals Participating	Approx. Weight of Float	Throwing Candy?
4-H	8	24	1.5 Tons	No
Larson's Tires	4	0	1.75 Tons	Yes
ABC Daycare	12	2	1.5 Tons	Yes
Central High Marching Band	6 (56 marching)	0	2.25 Tons	No
Bob's Feed and Grain	8	2	3 Tons	Yes
Girl Scouts	24	4	1.6 Tons	Yes
Boy Scouts	18	0	1.5 Tons	Yes
Central Farm Co-op	6	2	2.25 Tons	Yes
Acme Hardware	4	1	2 Tons	No

Solve.

1 Helen wants to arrange the floats so that the smallest floats are at the front and the largest are at the back. Assume that the weight of the float indicates its relative size. How would you order the floats?

2 The judges ask Helen to tell them the average number of animals participating. She thinks they really want to know the most common number of animals on the floats. Calculate the average, and then calculate the mode. Then write a sentence telling why Helen thinks the mode will be more useful than the mean.

3 The parade committee has to pay a special tax to the city if the average weight of the floats is over 2 tons. Will they have to pay this tax this year?

4 What is the range of people riding on the floats? (Do not include the 56 members of the marching band, only the six members of Homecoming Court riding with them.)

Bar Graphs

DATA ANALYSIS AND PROBABILITY

Enrollment for City Parks and Recreation Courses, Spring

Weaving: 8	Painting: 8	Knitting: 6
Tap Dance: 12	Aerobics: 48	

1. Determine the range of the data. Use this to set up and label the rows of your graph.

2. Determine the categories for your bars. Use this to set up your columns and label each category.

3. Give the graph a title.

Graph the data.

_____

Standards–Based Math • 5–6 © 2004 Creative Teaching Press

Name _____ Date _____

Pie Graphs

DATA ANALYSIS AND PROBABILITY

Complete the chart. Calculate the total number of apples sold first. Then use the data to create a pie graph.

Apple Type	Number Sold	Percent of Whole
Red Delicious	200	
Golden Delicious	350	
Granny Smith	175	
Jonathan	50	
Winesap	55	
McIntosh	70	

Find the degrees of each pie wedge by multiplying the percentage of the circle times 360°.

McIntosh	70	8%

$$0.08 \times 360° = 29°$$

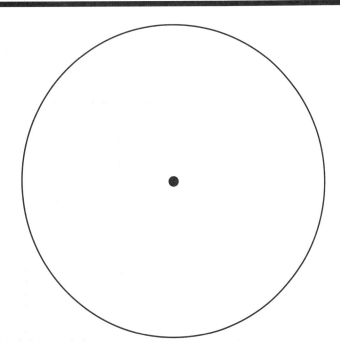

Name _____ Date _____

Line Graphs

DATA ANALYSIS AND PROBABILITY

Plot the data on the line graph. Then use it to answer the questions.

Distance Traveled Over Time

Time (seconds)	Distance (Meters)
0	0
1	2
2	8
3	18
4	32
5	50
6	72
7	98
8	128
9	162
10	200

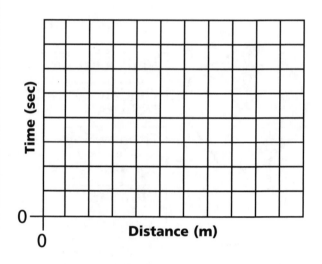

1 What is the range of the data?

2 How far did the object travel from 4 seconds to 6 seconds?

3 Did the object ever reverse direction?

Standards–Based Math • 5–6 © 2004 Creative Teaching Press

Probability and Statistics Vocabulary

DATA ANALYSIS AND PROBABILITY

Write the word that best completes each clue. Use these words to complete the crossword puzzle.

Across

5. _____ is the measurement of how likely it is that an event will occur.

Down

1. The _____ is a the method of collecting information.

2. An educated guess at an answer to a question is a _____.

3. The _____ is a set of data gathered.

4. The most basic outcome of an experiment is the _____.

5. The entire group or set from which you want to gather data is the _____.

Determining Probability

DATA ANALYSIS AND PROBABILITY

Probability = $\dfrac{\text{no. of events you want}}{\text{total possible events}}$

There are 20 marbles in the bag.

3 marbles are **green**.
6 marbles are **yellow**.
1 marble is **black**.
5 marbles are **purple**.
The remaining marbles are purple and gold swirled together.

After you pick a marble, you look at the color, note it, and then put the marble back in the bag so that each time you pick there are always 20 marbles.

Read and solve each problem.

1 What is the probability that you will pick a solid yellow marble?

2 What is the probability that you will pick a solid green marble?

3 What is the probability that you will pick a swirled marble?

4 What is the probability that you will pick a solid black marble?

5 What is the probability that you will pick a solid purple marble?

6 What is the probability that you will pick any solid marble?

7 What is the probability that you will pick a solid red marble?

8 What is the probability that you will pick a marble with gold on it?

9 What is the probability that you will pick a yellow or green marble?

10 What is the probability that you will pick a marble with any purple on it?

Standards–Based Math • 5–6 © 2004 Creative Teaching Press

Name _____ Date _____

Certain or Impossible?

Data Analysis and Probability

What is the probability of spinning a 5? It is impossible to spin a 5 because the spinner does not contain this number. 5 is not a possible simple event.

The opposite of an impossible event is a certain event. It is certain you will spin a 5 on a spinner that only contains the number 5.

Read each situation and write **certain** or **impossible**.

1 You are playing Add 'Em Up Domino with your 3-year-old brother using regular 9-dot dominos. The rules are simple. Each player earns the number of points on the open end of each domino he or she places on the board. Your brother is pretty good at matching the dots but is still a little fuzzy with his number identification. He places one domino and announces that he gets "20." How likely is this?

_____

2 Now your brother is thirsty. You take him to the kitchen and he announces he wants a juice box. All the juice boxes are apple juice. Sticking out his lower lip, he places his hands on his hips and announces authoritatively, "I only want the apple ones!" You laugh. Are you laughing because his choice is certain or impossible?

_____

3 Surprise! You open your closet this morning to find that your beloved cat has given birth to seven kittens! A closer examination reveals that all the kittens are males. Your uncle offers to take three of the kittens to work as mousers on his farm. What is the probability he will choose a boy kitten?

_____

4 You choose a number from the following list: 2, 3, 5, 7, 11, 13, 17, 19. What is the probability that the number will be a composite number?

_____

Answer Key

Number Value (Page 4)

1. four hundred sixty-two
2. four hundred two
3. four hundred sixty and three tenths
4. eighty-two thousand nine hundred seventy-four
5. fifty thousand four and eight hundredths
6. four million seven hundred sixty-eight thousand three hundred twelve
7. seven hundred
8. three hundred thousand
9. five thousandths
10. thirty
11. 645; 564; 546; 465
12. 5,043; 4,503; 3,708; 3,087
13. 44.03; 43.04; 40.34; 34.04
14. 345.54; 254.45; 234.98; 234.89

Number Notations (Page 5)

1. $300 + 60 + 5$; $(3 \times 10^2) + (6 \times 10) + 5$
2. $9,000 + 700 + 80 + 7$; $(9 \times 10^3) + (7 \times 10^2) + (8 \times 10) + 7$
3. $3,000 + 700$; $(3 \times 10^3) + (7 \times 10^2)$
4. $40,000 + 2,000 + 800 + 90 + 9$; $(4 \times 10^4) + (2 \times 10^3) + (8 \times 10^2) + (9 \times 10) + 9$
5. $5,000,000 + 50,000 + 500 + 50 + 5$; $(5 \times 10^6) + (5 \times 10^4) + (5 \times 10^2) + (5 \times 10) + 5$
6. $200 + 90 + 1$
7. $(5 \times 10^3) + (7 \times 10^2) + (9 \times 10) + 8$
8. $(7 \times 10^3) + (8 \times 10)$
9. $40,000 + 3,000 + 200 + 50 + 6$
10. $30,000 + 3,000 + 20 + 4$

Inverse Operations (Page 6)

1. 808
2. 4,589
3. 187
4. 142,411
5. 88,855
6. 1,768
7. 83,010
8. 21,517
9. 297,834
10. 110,932
11. 35,108
12. 111,203

Riddle (Page 7)

1. 16,961

2. 5,080
3. 153
4. 136,233
5. 85,091
6. 16,958
7. 147,200
8. 22,189
9. 287,617
10. 129,878
11. 20,389
12. 163,695
THE LETTER G.

Comparing Integers (Page 8)

1. 7, -9
2. 12, -12
3. 0, -8
4. -2, -13
5. $1" + 2" + - 4" = -1"$
6. $\$30 + -\$10 + -\$15 + \$10 = \$15$
7. $-\$1,000 + \$200 = -\$800$

Subtracting Integers (Page 9)

1. $-72 + (-70) = -142$
2. $52 + 46 = 98$
3. $90 + (-49) = 41$
4. $7 + 95 = 102$
5. $-55 + 59 = 4$
6. $-47 + (-20) = - 67$
7. -13
8. 56
9. -145
10. -57
11. -3
12. 24
13. -2
14. 141
15. 49
16. 30

Multiplying Integers (Page 10)

1. -21 -21 21 21
2. -30 -30 30 30
3. -27 -27 27 27
4. -18 -72 35 12
5. 14 -32 -24 36
6. 40 -56 54 -15
7. -80 96 84 72

8. -84 -180 -72 36
9. -147 -100 80 252

Dividing Integers (Page 11)

1. -7 -7 7 7
2. -2 -2 2 2
3. -12 -12 12 12
4. -8 5 -3 9
5. -4 6 8 -11
6. 3 -5 -7 -12
7. -5 13 14 -36
8. 16 25 -15 -21
9. 31 -5 42 -51

Multiplying and Dividing Integers (Page 12)

1. -168 13 -600 22
2. 516 -38 -180 28
3. 371 192 14 6
4. -24 216 -47 -228
5. -204 20 46 -252
6. -396 14 -203 32
7. -15 245 343 -26
8. 18 -120 120 -75
9. 7 252 16 -256

Checks and Balances (Page 13)

1. $920.50
2. $670.50
3. $590.50
4. $234.64
5. $158.99
6. $198.99
7. $126.34
8. ‑$163.12
9. ‑$198.35
10. $551.65
11. $1,751.65
12. $1,501.65
13. $1,235.67
14. $870.67

Mental Math with Zeros (Page 14)

1. 28,000
2. 6
3. 320,000
4. 30
5. 1,800,000
6. 20
7. 1,800
8. 7
9. 4,800,000

10. 70
11. 720,000
12. 9
13. 350,000
14. 40

Multiplying Larger Numbers (Page 15)

1. 266,175
2. 181,712
3. 815,442
4. 293,094
5. 139, 620
6. 222,075
7. 172,082
8. 277,119
9. 514,230
10. 571,032
11. 563,103
12. 51,597

Fraction Review (Page 16)

1. 1/2
2. 1/4
3. 2/6
4. 2/3
5. 2/5
6. 1/6
7. 2/3
8. 2/6
9. 1/2

Fraction of a Set (Page 17)

1. 2/3; 1/3
2. 2/5; 3/5
3. 3/4; 1/4
4. 3/6 or 1/2; 3/6 or 1/2
5. 2/6 or 1/3; 4/6 or 2/3
6. 3/10; 7/10

Converting Fractions to Decimals (Page 18)

1. 0.2
2. 0.25
3. 0.3
4. 0.4
5. 0.05
6. 0.05
7. 0.8

8. 0.8
9. 0.1
10. 1
11. 0.625
12. 0.4
13. 0.15
14. 0.08

Converting Fractions to Repeating Decimals (Page 19)

1. $0.\overline{33}$
2. $0.2\overline{77}$
3. $0.\overline{33}$
4. $0.\overline{55}$
5. $0.\overline{190476}$
6. $0.20\overline{833}$
7. $0.\overline{259}$
8. $0.1\overline{2}$
9. $0.\overline{33}$
10. $0.0\overline{925}$

Converting Decimals to Percents (Page 20)

1. 20%
2. 25%
3. 30%
4. 40%
5. 5%
6. 5%
7. 80%
8. 80%
9. 10%
10. 100%
11. 62.5%
12. 40%
13. 15%
14. 8%

Approximate Percents (Page 21)

1. 33%
2. 28%
3. 33%
4. 56%
5. 19%
6. 21%
7. 26%
8. 12%
9. 33%
10. 9%

Multiples (Page 22)

1. 7: 7, 14, 21, 28, 35, 42, 49, 56, 63, 70
2. 12: 12, 24, 36, 48, 60, 72, 84, 96, 108, 120
3. 3: 3, 6, 9, 12, 15, 18, 21, 24, 27, 30
4. 9: 9, 18, 27, 36, 45, 54, 63, 72, 81, 90
5. 4: 4, 8, 12, 16, 20, 24, 28, 32, 36, 40
6. 11: 11, 22, 33, 44, 55, 66, 77, 88, 99, 110
7. 8: 8, 16, 24, 32, 40, 48, 56, 64, 72, 80
8. 5: 5, 10, 15, 20, 25, 30, 35, 40, 45, 50
9. 10: 10, 20, 30, 40, 50, 60, 70, 80, 90, 100
10. 2: 2, 4, 6, 8, 10, 12, 14, 16, 18, 20
11. 6: 6, 12, (18) 24, 30, (36) 42, 48, (54) 60
    9: 9, (18) 27, (36) 45, (54) 63, 72, 81, 90
12. 4: 4, 8, (12) 16, 20, (24) 28, 32, (36) 40
    12: (12) (24) (36) 48, 60, 72, 84, 96, 108, 120

Least Common Multiple (Page 23)

1. 5: 5, 10, 15, 20, 25, 30, 35, (40)
    8: 8, 16, 24, 32, (40)

2. 7: 7, 14, 21, 28, (35)
    5: 5, 10, 15, 20, 25, 30, (35)

3. 3: 3, 6, 9, 12, 15, 18, 21, (24)
    8: 8, 16, (24)

4. 9: 9, 18, 27, (36)
    12: 12, 24, (36)
    6: 6, 12, 18, 24, 30, (36)

5. 15: 15, 30, 45, (60)
    10: 10, 20, 30, 40, 50, (60)
    20: 20, 40, (60)

6. 12: 12, 24, 36, (48)
    16: 16, 32, (48)

Greatest Common Factor (Page 24)

1. 24: 1, 2, 3, 4, 6, 8, (12) 24
    36: 1, 2, 3, 4, 6, 9, (12) 18, 36

2. 36: 1, 2, 3, 4, 6, (9) 12, 18, 36
    9: 1, 3, (9)

3. 21: 1, 3, (7) 21
    28: 1, 2, 4, (7) 14, 28

4. 32: 1, 2, 4, (8) 16, 32
    40: 1, 2, 4, 5, (8) 10, 20, 40
    16: 1, 2, 4, (8) 16,

5. 45: 1, 3, 5, 9, (15), 45
   15: 1, (15)
   60: 1, 2, 3, 4, 5, 6, 10, 12, (15), 20, 30, 60

6. 12: 1, 2, 3, (4), 6, 12
   16: 1, 2, (4), 8, 16

Reducing Fractions (Page 25)

1. 2/21
2. 3/5
3. 2/5
4. 3/4
5. 3/4
6. 3/4
7. 1/3
8. 2/3
9. 1/6
10. 5/8
11. 2/3
12. 9/16
13. 1/3
14. 1/8
15. 1/3
16. 1/5

Adding Fractions (Page 26)

1. 11/12
2. 9/11
3. 4/5
4. 11/12
5. 15/22
6. 23/24
7. 13/15
8. 9/10
9. 7/8
10. 5/12
11. 11/12
12. 23/24

Subtracting Fractions (Page 27)

1. 1/6
2. 5/11
3. 2/3
4. 1/12
5. 4/11
6. 5/24
7. 1/3
8. 1/2
9. 11/24
10. 1/3
11. 7/90
12. 1/12

Improper Fractions to Mixed Numbers (Page 28)

1. 1 2/3
2. 3 3/4
3. 3 1/9
4. 2 3/7
5. 6 1/2
6. 6 4/5
7. 3 6/11
8. 4 2/3
9. 5 2/3
10. 1 11/19
11. 6 13/19
12. 5 3/4
13. 4 2/5
14. 2 1/6
15. 4 4/5
16. 4 5/11
17. 3 5/8
18. 5 1/6

Mixed Numbers to Improper Fractions (Page 29)

1. 54/7
2. 45/11
3. 57/5
4. 74/9
5. 63/10
6. 66/8
7. 77/6
8. 75/6
9. 78/9
10. 73/8
11. 76/12
12. 64/5
13. 67/8
14. 55/6
15. 46/5
16. 56/5
17. 79/11
18. 65/6

Multiplying Fractions (Page 30)

1. 1/7
2. 21/32
3. 2/15
4. 1/4
5. 10/27
6. 1/7
7. 3/8
8. 2/7
9. 4/21
10. 4/9
11. 2/7
12. 3/20
13. 5/49
14. 3/25
15. 15/56
16. 1/14

Multiplying Mixed Numbers (Page 31)

1. 23 7/24
2. 12 2/3
3. 15 5/12
4. 22 27/32
5. 6 8/9
6. 26 1/4
7. 35 7/15
8. 8 6/7

Dividing Fractions (Page 32)

1. 2 5/8
2. 2/9
3. 5/7
4. 5/6
5. 27/35
6. 21/32
7. 1 2/3
8. 7/10
9. 3/5
10. 16/21
11. 1 2/3
12. 3/4
13. 1 2/7
14. 7/24
15. 25/36

Decimals to Fractions (Page 33)

1. 4 1/4
2. 12/25
3. 3 7/10
4. 1/250
5. 4 33/250
6. 9/100
7. 1 3/4
8. 1/4
9. 2/5
10. 2 4/5

Percent of a Set (Page 34)

1. 66%
2. 50%
3. 33%
4. 83%
5. 25%
6. 10%

7.
8.
9.
10.
11.
12.

Percent of a Value (Page 35)

1. 3.96
2. 245
3. 750
4. 5.4
5. 16
6. 24mL
7. 5m
8. $138
9. 22.8
10. 95.04
11. $15
12. 108
13. 100
14. 20
15. $8.00
16. 15

Leaving a Tip (Page 36)

1. 98¢
2. $6.75
3. $3.43
4. $2.76
5. $1.18
6. $2.13
7. $8.00
8. $7.14
9. $4.40
10. $1.54

Paying Sales Tax (Page 37)

1. $1.10
2. $0.47
3. $0.84
4. $0.79
5. $1.27
6. $10.20
7. $5.05
8. $2.26
9. $15.89

10. $16.75
11. $14.57
12. $16.61
13. $72.34
14. $34.92

Analogies 1 (Page 38)

1. c
2. e
3. d
4. a
5. b
6. g
7. f

Analogies 2 (Page 39)

1. b
2. a
3. c
4. c
5. a
6. b

Patterns (Page 40)

1.

2.

3.

4.

5.

6.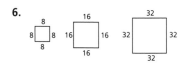

Number Patterns 1 (Page 41)

1. × 5
2. + 14
3. ÷ 3
4. × 3
5. -7
6. ÷ 6
7. × 12
8. -2
9. + 8
10. × 6
11. + 31
12. ÷ 8

Number Patterns 2 (Page 42)

1. 13, 15, 17
2. 65, 60, 55
3. 21, 19, 24
4. 108, 126, 144
5. 49, 56, 63
6. 60, 70, 80
7. 80, 110, 145
8. 33, 40, 47
9. 37, 46, 55
10. 29, 31, 37

Order of Operations (Page 43)

1. 19
2. 4
3. 64
4. 5
5. 8
6. 8
7. 125
8. 61
9. 50
10. 7
11. 4
12. 56
13. 3
14. 12

Commutative Property (Page 44)

×	2	4	$\frac{1}{2}$	-3	5
$\frac{1}{3}$	$\frac{2}{3}$	$1\frac{1}{3}$	$\frac{1}{6}$	-1	$1\frac{2}{3}$
6	12	24	3	-18	30
$\frac{3}{5}$	$1\frac{1}{5}$	$2\frac{2}{5}$	$\frac{3}{10}$	$-1\frac{4}{5}$	3
5	10	20	$2\frac{1}{2}$	-15	25
8	16	34	4	-24	40

2 × 1/3; 1/3 × 2
2 × 6; 6 × 2
2 × 3/5; 3/5 × 2
2 × 5; 5 × 2
2 × 8; 8 × 2
4 × 1/3; 1/3 × 4
4 × 6; 6 × 4
4 × 3/5; 3/5 × 4
4 × 5; 5 × 4
4 × 8; 8 × 4
1/2 × 1/3; 1/3 × 1/2
1/2 × 6; 6 × 1/2
1/2 × 3/5; 3/5 × 1/2
1/2 × 5; 5 × 1/2
1/2 × 8; 8 × 1/2
-3 × 1/3; 1/3 × -3
-3 × 6; 6 × -3
-3 × 3/5; 3/5 × -3
-3 × 5; 5 × -3
-3 × 8; 8 × -3
5 × 1/3; 1/3 × 5
5 × 6; 6 × 5
5 × 3/5; 3/5 × 5
5 × 5; 5 × 5
5 × 8; 8 × 5

Associative Property (Page 45)

1. 48	**2.** 19
3. 60	**4.** 23
5. 42	**6.** 60
7. 42	**8.** 109
9. 44	**10.** 60
11. 969	**12.** 48
13. 379	**14.** 108

Distributive Property (Page 46)

1. 21	**2.** -32
3. 22	**4.** -10
5. 56	**6.** 45
7. -32	**8.** 33
9. 28	

Exponents (Page 47)

1. 125
2. 16
3. 81
4. 36
5. 512
6. 81
7. 625

8. 343
9. 64
10. 1,000
11. 144
12. 1,331

Exponential Notation (Page 48)

1. 1,000
2. 100,000,000
3. 10,000,000
4. 720,000
5. 60,000,000
6. 0.00087
7. 9,000,000,000
8. 0.00000002
9. 0.000001204
10. 0.0000006
11. 2,000,000,000,000
12. 11,100,000,000

Variables (Page 49)

1. 3 + n = 8
2. n + 200 = 1,000
3. 2 + x = 7
4. x + 12 = 48

Inverse Operations (Page 50)

1. e	**2.** g
3. j	**4.** f
5. c	**6.** k
7. b	**8.** d
9. l	**10.** i
11. a	**12.** h

Inverse Operations: Addition (Page 51)

1. x = 8
2. a = 27
3. y = 13
4. z = 42
5. x = 10
6. a = 85
7. x = 85
8. a = 18
9. y = 59
10. b = 36
11. x = 20
12. c = 0

Inverse Operations: Subtraction (Page 52)

1. x = 16
2. a = 30
3. y = 25
4. z = 95
5. x = 85
6. a = 145
7. x = 115
8. a = 33
9. y = 22
10. b = 6.4
11. x = 4
12. c = 4

Inverse Operations: Multiplication (Page 53)

1. x = 5
2. x = 10
3. a = 9
4. x = 20
5. c = 4
6. x = 6
7. b = 7
8. n = 7
9. x = 9
10. x = 15
11. a = 100
12. x = 5

Inverse Operations: Division (Page 54)

1. a = 42
2. x = 48
3. x = 48
4. b = 45
5. x = 20
6. n = 120
7. x = 27
8. x = 75
9. c = 6
10. x = 88
11. x = 25
12. a = 55

Variables 1 (Page 55)

HELEN OF TROY

Variables 2 (Page 56)

ALEXANDER THE GREAT

Variables 3 (Page 57)

MISSOURI IN 1800S

Variables 4 (Page 58)

ROADS AND HOUSES

Evaluating Expressions (Page 59)

1. 10
2. 1 1/3
3. 29
4. 8
5. 6
6. 53
7. 9/20
8. -81
9. 3
10. 126
11. x = 4
12. c = 4

Inequalities (Page 60)

1. < 3
2. < -144
3. < -1 1/3
4. > 61
5. < 194
6. > 1
7. < 2/3
8. > -55
9. > 1
10. > -180
11. > 1/8
12. > 16

Functions 1 (Page 61)

$3x = y$ y   6   12   18   24   30   36

$4x - 8 = y$ y   4   16   28   40   52   64

$3x/6 = y$ y   1/2   1 1/2   2 1/2   3 1/2   4 1/2   5 1/2

$5x + 3 = y$ y   13   18   23   28   33   38

$10^x = y$ y   0.001   0.01   0.1   1   100   10,000

$x+2/x = y$ y   1 1/4   1 2/9   1 1/5   1 2/11   1 1/6   1 2/13

Functions 2 (Page 62)

$3(x + 4) = y$ y   18   24   30   36   42   48

$4(x − 8) = y$ y   -20   -8   4   16   28   40

$4x/8 = y$ y   1/2   1 1/2   2 1/2   3 1/2   4 1/2   5 1/2

$6x + 7 = y$ y   19   25   31   37   43   49

$10^x = y$ y   0.00001   0.0001   0.001   10   1,000   100,000

$x + 5/x = y$ y   1 5/8   1 5/9   1 1/2   1 5/11   1 5/12   1 5/13

Functions 3 (Page 63)

$x/6 = y$ y   1/3   2/3   1   1 1/3   1 2/3   2

$0.6x + 0.6 = y$ y   2.4   4.2   6   7.8   9.6   11.4

$-5/3x + 4 = y$ y   2 1/3   -1   -4 1/3   -7 2/3   -11   -14 1/3

$-1.5x + 2.5 = y$ y   -0.5   -2   -3.5   -5   -6.5   -8

$4x + 8 = y$ y   -12   -8   -4   12   20   28

$2/3x + 1/3 = y$ y   5 2/3   6 1/3   7   7 2/3   8 1/3   9

Writing and Solving Equations (Page 64)

1. $6n = 60$ or $60/6 = n$; $n = 10$
2. $144/n = 12$ or $12n = 144$; $n = 12$
3. $16 + n = 22$ or $22 − 16 = n$; $n = 6$
4. $5000/n = 250$ or $250n = 5000$; $n = 20$
5. $4n = 1,660$ or $1,660/4 = n$; $n = 415$
6. $22 + n > 40$; $n > 17$

Venn Diagrams (Page 65)

1. $n = 3$
2. $n = 6$

Angles (Page 66)

1. b
2. f
3. d
4. a
5. e
6. c

7.
8.
9.

Angle Measurement (Page 67)

1. 60°
2. 47°
3. 25°
4. 15°
5. 45°
6. 142°

Geometry Terms (Page 68)

1. intersecting
2. plane
3. angle
4. face
5. midpoint
6. point, perpendicular
7. ray
8. edge
9. solid
10. parallel

Quadrilaterals (Page 69)

1. 90°; square
2. 140°; parallelogram
3. 90°; rectangle
4. 95°; rhombus
5. 60°; trapezoid
6. 41°; parallelogram

Polygons (Page 70)

1. nonagon
2. pentagon
3. decagon
4. quadrilateral
5. hexagon
6. heptagon
7. triangle
8. octagon
9. quadrilateral
10. quadrilateral
11. dodecagon
12. dodecagon

Similar and Congruent (Page 71)

1. similar
2. congruent
3. congruent
4. similar
5. congruent
6. congruent
7. similar
8. similar
9. congruent

Similar Triangles (Page 72)

1.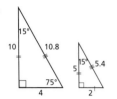

2.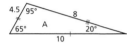

Similar Figures (Page 73)

1.

2.

3.

4.

Prisms and Pyramids (Page 74)

1. prism; octagon
2. pyramid; rectangle
3. prism; triangle
4. prism; rectangle
5. pyramid; pentagon
6. pyramid; square

Cylinders, Cones, and Spheres (Page 75)

1. cone
2. sphere
3. cylinder
4. pyramid
5. sphere
6. pyramid
7. prism
8. prism
9. cylinder

Pythagorean Proof (Page 76)

Pythagorean Theorem (Page 77)

1. 5.657
2. 9.434
3. 10.44
4. 47.424
5. 60.415
6. 118.983
7. 3.905
8. 72.56
9. 68.352
10. 14.422

Parts of a Circle (Page 78)

1. c
2. e
3. d
4. a
5. f
6. b

Possible answers include:

The X, Y Axis (Page 79)

1. (2, 3)
2. (2, −2)
3. (4, 2)
4. (−4, 1)

Coordinate Pairs (Page 80)

1. (–8, 9)
2. (–8, 5)
3. (–5, 8)
4. (3, –5)
5. (7, –4)
6. (7, 1)
7. (3, 2)
8. (1, –2)
9. (–6, –1)
10. (–4, –4)
11. (–4, –1)
12. (1, 4)
13. (3, 4)
14. (3, 7)

Using a Cartesian Plane (Page 81)

1. (–1, 3) 2. (–1, 4)
3. (–2, 3) 4. (5, 9)
5. (–2, –9) 6. (6, 0)
7. (–3, 4) 8. (–6, 11)

Plotting Points (Page 82)

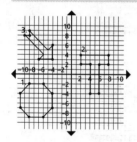

Reflection Symmetry (Page 83)

1. yes
2. no
3. yes
4. yes
5. no
6. yes

7.   8.   9.

Reflection Symmetry 2 (Page 84)

Students need only show one line for each figure.

1.

2.

3.

4.

5.

6.

7.

8.

9.

10.

11.

12.

Rotational Symmetry (Page 85)

1. yes 2. yes
3. yes 4. yes
5. yes 6. no
7. no 8. no
9. yes 10. yes
11. yes 12. yes

Reflections Across the X-Axis (Page 86)

1. Reflection (2, 8)

(2, 2)

(4, 2)

(4, 6)

(7, 6)

(7, 8)

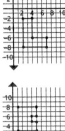

2. Reflection (1, −3)

(1, −8)

(5, −8)

(5, −6)

(4, −6)

(4, −5)

(5, −5)

(5, −3)

3. Reflection (−3, −3)

(−5, −6)

(−3, −9)

(−5, −9)

(−6, −7)

(−7, −9)

(−9, −9)

(−7, −6)

(−9, −3)

(−7, −3)

(−6, −5)

(−5, −3)

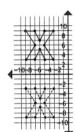

4. Reflection (−6, 2)

(−4, 4)

(−6, 5)

(−4, 8)

(−8, 8)

(−10, 5)

(−8, 4)

(−10, 2)

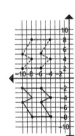

Reflections Across the Y-Axis (Page 87)

1. Reflection (2, 6)

(2, 8)

(5, 8)

(5, 7)

(7, 7)

(7, 4)

(5, 4)

(5, 6)

2. Reflection (3, 3)

(3, 6)

(4, 5)

(4, 8)

(7, 7)

(7, 5)

(6, 6)

(6, 3)

(7, 2)

(4, 2)

3. Reflection (−2, 4)

(−2, 6)

(−4, 9)

(−5, 7)

(−8, 9)

(−8, 6)

(−5, 6)

(−6, 4)

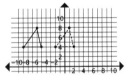

4. Reflection (−4, 5)

(−6, 7)

(−9, 7)

(−11, 5)

(−8, 5)

(−10, −2)

(−5, −2)

(−7, 5)

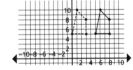

Translations Along the X-Axis (Page 88)

1. Translation (−2, 4)

(1, 8)

(2, 4)

2. Translation (10, 5)

(11, 10)

(13, 8)

(13, 5)

3. Translation (−1, −9)

(3, −3)

(4, −6)

(2, −9)

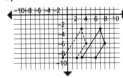

4. Translation (2, 2)

(1, 4)

(2, 6)

(1, 8)

(2, 10)

(4, 10)

(3, 8)

(4, 6)

(3, 4)

(4, 2)

Translations Along the Y-Axis (Page 89)

1. Translation (2, 0)
(2, 2)
(8, 0)
(7, –1)
(4, –1)

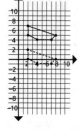

2. Translation (–10, 3)
(–9, –1)
(–7, 0)
(–5, 2)
(–8, 2)

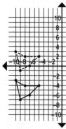

3. Translation (3, –5)
(3, –3)
(5, –2)
(7, –3)
(9, –2)
(9, –5)

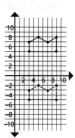

4. Translation (–10, 6)
(–7, 6)
(–5, 4)
(–3, 6)
(–3, 1)
(–5, 3)
(–7, 1)
(–10, 1)

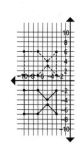

Oblique Translations (Page 90)

1. Translation (7, –1)
(7, 0)
(11, 0)
(11, –1)

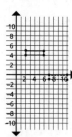

2. Translation (–6, 9)
(–6, 13)
(–4, 11)
(–4, 8)

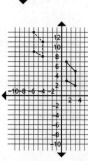

3. Translation (9, –5)
(6, –7)
(9, –8)
(12, –7)

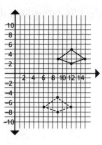

4. Translation (4, 6)
(6, 4)
(5, 3)
(6, 1)
(4, 1)
(3, 3)
(4, 4)
(2, 6)

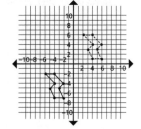

Transformations (Page 91)

1. reflection across y-axis or translation along x-axis
2. reflection across x-axis
3. reflection across y-axis
4. oblique translation
5. translation along y-axis
6. oblique translation

Perimeter (Page 92)

1. 16"   **2.** 22 ft.
3. 48 m   **4.** 45 cm
5. 40   **6.** 32 km
7. 60   **8.** 36
9. 80

Area (Page 93)

1. 20 sq. units
2. 18 sq. units
3. 31 sq. units
4. 42 sq. units
5. 34 sq. units
6. 36 sq. units

Area of a Triangle (Page 94)

1. 200
2. 120
3. 32 sq. cm
4. 600
5. 18
6. 10 sq. in.